Suppression

P.S. Winn

© 2014

Although actually based on a true story, this book is a work of fiction. Names, places, characters and happenings are a work of the author's imagination. Any resemblance to people, places or actual happenings is purely coincidental.

4

Chapter 1

Looking up as her door opened, Mary turned from her place by the small window and frowned. She didn't get many visitors and the nurses had already been by. Mary shook her head slightly as she stood and then stepped toward Ned who was standing in the doorway.

Ned was an orderly here at Rose Hills and in one sentence, 'a complete ass'. Mary didn't trust him and his too many questions and then there were those know it all sneers.

Mary had been in the psychiatric hospital for six months now and wasn't expecting to get out anytime soon either.

Mary's blue eyes widened as Ned stepped to one side and a handsome blonde haired man stepped in to her room. He flashed Mary a brilliant smile which made his green eyes light up.

Ned, a dumpy five foot ten was a couple inches shorter and paled in comparison to the young man. Ned was staring at Mary with a hard glare in his dark eyes.

"Mrs. Delany has agreed to let Mr. Franklin here visit with you a few hours a day. He's a reporter."
Ned almost spit out the last word before adding.
"He's interested in your story."

The man stepped forward and reached out a hand, adjusting the bag he had slung over his shoulder as he did. Mary took the man's hand in her.

The man smiled. "Hi, I'm Chad Franklin, and I am really glad to meet you. I work at 'The Vicinage Voice'." He laughed. "I know, dumb name, but my boss likes it and he's the one who signs the checks. Anyway, he has agreed to let me do a story about you and your husband and that invention of his."

Mary frowned and her blue eyes got glassy. "My husband is dead."

Chad nodded. "I know, and I am very sorry. That is just another reason his story should be told. If you aren't ready to talk about it, I'll understand, but I am really hoping you will be."

Mary studied the man in front of her. He was probably in his mid-twenties, almost ten years younger than she was. His face looked so animated, that she felt her own spirits rise. She glanced over at Ned and it was like a dark cloud had passed over. She turned back to Chad, much better.

Mary smiled. "I think I would like to talk to you Mr. Franklin."

Chad smiled. "That's great and just call me Chad, my dad is Mr. Franklin."

Mary gave him a slight smile and Chad realized how pretty she was. He hoped he would be able to get her to smile more often over the next week or however long it took to hear her story. Chad looked around the room they stood in and could understand why Mary didn't smile much. The room was drab and a little dingy with no personal touches. He turned to look at Ned and thought this guy with his negative attitude probably didn't help much either.

"You probably have work to do. Mary and I will be fine."

Glaring at Chad, Ned turned and stomped out of the room. As soon as he was gone, Chad sighed. "Guess he didn't like that too well."

Mary shrugged. "That's just Ned, he's always like that."

Mary pointed to a small table that had three mismatched chairs sitting around it.
"Why don't we go ahead and sit over there?"

Chad slid the bag off of his shoulder and set it on the table as he sat down.
Mary took a seat opposite of him. "How come they let you in? I don't get many visitors."
Mary laughed and pointed at her own head.
"I'm crazy you know."

Chad shook his head. "I don't think so and neither does my editor or I wouldn't be here."

Mary half shrugged. "According to my Doctor and a few

Senators who hold him in their pockets, I am more than just a little crazy."

Leaning forward, Mary looked at Chad's face, studying it; finally she nodded, liking what she saw.

"It was the Government who killed my husband. They killed Dan because of his invention, if you don't believe that you probably don't want to hear my story or rather our story. This whole thing is more about Dan than anything, it is his story."

Chad's face turned serious, his green eyes a slightly darker shade than they had been earlier.
"I believe you Mary. That's why I want to hear your story. Why this story has to be told so others can hear it too."

Mary nodded. "Aren't you afraid? Dan's dead and I'm in this nut house. Telling this story might not be the best or safest thing to do. The people suppressing all of this don't play little kids games, they are serious and deadly."

Chad looked at her. "I think you're wrong Mary. I think this is a story that needs to be told and to the biggest audience possible."

Mary thought about that a minute and then nodded. "If you're willing to hear it out, then I think I am ready to tell it."

Chad nodded and opened the bag in front of him. He pulled out a pen and a notebook.
He reached back in the bag and pulled out a small tape recorder and looked at Mary.
"I'd like to record you. I'll be taking notes, but I don't want to miss anything. Is that okay?"

Looking warily at the recorder, Mary shrugged. "Yeah, I guess I'm okay with that."

Chad smiled. "Good."

Mary looked at him and Chad could see the dark circles under her blue eyes. "Where do you want me to start?"

Chad shrugged as he stared at the face in front of him. "Anywhere you want. When did Dan start on the invention?"

Mary smiled. "Dan was always working on one invention or another. He was a fixer."

Mary's voice cracked just slightly, but she took a second, cleared her throat and kept talking.

"Dan was the kind of person who could look at something broken and figure a way to fix it. He was the same way with people. He'd do anything for anybody. In fact I used to get mad at him, tell him that nice guys finished last."

Mary shook her head. "Didn't matter, he was a giver with a heart of gold. Even though those bastards killed him for it, I'm pretty sure if he had a second chance he'd do it all the same way. He would have made things better for the whole world if it wasn't for those people with power and money stopping him."

Half smiling, Mary shrugged her shoulders.

"But you want to hear about his last invention."

Chad nodded. "Yeah, but I wouldn't mind after we finish with that hearing about the others. Dan Warren must have been an amazing man."

Swallowing the lump in her throat, Mary nodded. "Yes, he was." She took a deep breath. "I guess I should start back when he first told me the idea. I'd say it started about ten years ago."

Chapter 2

Mary was at the stove cooking dinner when she heard the back door open. She looked at the clock and smiled. "I'm in the kitchen Dan."

Dan came walking in, at six foot two he filled up the room and was almost a foot taller than his wife; he had to lean over to kiss Mary on the cheek. The couple shared brown hair and blue eyes. Dan put down his lunchbox on the table and then turned to Mary. "How's your day honey?"

Mary smiled. "It's been good, better now that you're here."

Dan took a breath and closed his eyes.
"Something sure smells good. Why don't I go and get washed up?"

Mary stirred a pot on the stove. "You've got five minutes."

Ten minutes later they were seated across from each other. Dan looked at the food on the table and rubbed his hands together. He smiled broadly.
"I love your pork chops."

Mary smiled; no matter what she cooked, Dan would say he loved it.

She looked across at him. "How about you? How was your day?"

Dan's eyes widened. "Funny you should ask."

He took a bite of food, chewed and swallowed then continued. "At lunch today I was listening to the news and they were saying a lot of places now have to ship their garbage out on barges."

Mary frowned. "Then what?"

Dan shrugged. "Then nothing, it just floats around out there."

He leaned toward Mary and she could see the glint in his eye and she knew he had an idea.
"Now, if that garbage could be burned this kind of thing wouldn't have to happen."

Mary shook her head. "What about the pollution from burning the garbage?"

Dan sat back and smiled. "I think I know how to fix that. I want to try and build a small incinerator, with a stack that has a sprayer. If I can get the water up to temperature, run it through a coil and spray it out above the smoke the steam vapor produced will be small enough to capture the particulates in the smoke. Then I can contain them in the water as it collects in some kind of reservoir."

Mary thought for a minute. "What will you do with the water then?"

Dan smiled and nodded excitedly. "I think I got that figured out too. I'll have to draw up some ideas first and then build something so I can give it a try."

Mary laughed. "Well, finish your supper first."

Dan rolled his eyes and nodded. "No problem there."

*

A few nights later, Mary got out of bed and found Dan up late sitting at the kitchen table. "Honey, what are you doing?"

Dan shifted the papers he'd been working on.
"Sit down and I'll show you what I've got so far."

Mary sat down and Dan picked up a piece of paper with a drawing on it.

"Here's kind of what I am thinking."
He pointed at the drawing.

"The square on the bottom will be the incinerator. You'll be able to put garbage, tires or whatever you need to burn in there. The cylinder on top is the unit where the steam is going to be made."

Dan frowned. "I may have to preheat the water. I'll want steam pressure built up before any smoke comes up from the incinerator. At the top of the stack, the sprayers will force the steam out so it can capture the smoke.

When the force of the water and the captured particulates hit the side of the chamber it will bring everything down and carry it in to the containment units." Dan turned to Mary and smiled. "Of course that's simplified. Once I build a small, scale model you'll be able to see more of what I'm showing you here."

Mary nodded. "You know, something like this could really be big. If you could burn garbage and tires, that would be so important. Land pollution could be brought to a minimum without adding to the air pollution. If you are able to capture the smoke and particulates we could all breathe clean air again."

Dan nodded. "Let me build it and see if my idea even works first."

Mary smiled. "Okay, but I think we need to get back to sleep now. You still have to be to work in the morning."

Dan nodded. "Yeah, and tomorrow we have a big order of asphalt and I'll be running the hot plant and probably hauling it in the truck too."

Mary frowned. "How come you're doing both?"

Dan laughed. "When am I not?"

Chapter 3

Mary walked to the sink to turn on the water and jumped back when she felt a small shock. She shook her head. That was the third time the same thing had happened this morning.

Mary shrugged and walked outside where Dan had set up the invention he had built. It had taken him over three months, but he finally had the first proto-type built and ready for a trial run. Mary walked over to where Dan stood by the twelve foot tall invention. She had thought of something more like a table top thing when Dan had said a scale model. She supposed it was a scale model when compared to the large stacks being used. Mary stepped up to her husband. "Hey Dan, something is wrong in the kitchen. I keep getting shocked on the sink in there."

Dan turned from the knobs he had been adjusting to look at Mary. "It's probably just a short in the wiring somewhere. I'll check it out later. I want to give this thing a trial run first."

Dan smiled. "You can even help if you want."

Mary looked at the invention then at Dan and frowned. "I'd love to help, but I wouldn't have a clue where to start."

Dan laughed. "Don't worry; I'll walk you through it. The whole thing is pretty simple though. All the important stuff is inside. It will take care of itself. We just need to pump in the heated water and then get a fire going."

Mary nodded. "What are you burning?"

Dan looked at her. "I got some old rubber belting; I'll put that in there with some real pitchy wood."

Mary's eyes widened. "That's gonna make a hell of a lot of black smoke."

Dan laughed. "If this works like I think, it will only be black for a minute. Just keep your fingers crossed."

Mary held up crossed fingers. "I think it will work without that, but a little luck never hurt anyway."

Dan nodded. "That's for sure. Since I've preheated the water, I'm going to start pumping it in, then we'll start up the fire and see what happens."

Dan walked over to the stack and turned a few knobs. Mary could hear a hissing noise as the hot water turned to steam.

Dan turned to Mary and smiled. "Do you want to do the honors?" He held up a lighter.

Mary took it and opened the fire box that Dan had built. She lit his prepared fire. As the fire caught and grew,

black smoke poured out of the stack's open top.
Mary frowned, but Dan smiled.
"Just give it a minute."

Dan reached over to the table and grabbed a video
camera and handed it to Mary. "Could you film that
black smoke coming out? Keep filming; the invention
will be kicking in. It should only take a couple minutes
now."

Mary took the camera and stepped back about twenty
feet so she could capture all of the invention. As she
watched, the black smoke turned to gray and then almost
white and then nothing was coming out. Mary almost
dropped the camera in surprise at what she was seeing.
She yelled over at Dan.
"Can you open the door so I can get a picture of that fire
burning?"

Nodding, Dan leaned down and pulled open the metal
door. The flames filled the three foot square box. Mary
used the camera to zoom in on the fire and then back out.
Slowly she focused the camera lens from the fire box
and up the stack to the top where no smoke could be
seen.
She zoomed out again so the whole invention could be
seen with the raging fire below and no smoke on top. In
fact as she watched the top of the stack all that Mary
could see was a slight rippling of the air from the clean
steam and heat being emitted. Mary was smiling.
"I don't know how you did it Dan, but you sure did.
That fire should be pouring out black smoke and there's
nothing. This invention is going to change the world."

Dan laughed. "Wouldn't that be great? This world needs some changing."

Two hours later Dan shut down the incinerator and then got busy getting samples of the water he had collected. The last sample was black as pitch with all the contaminates captured. Dan put the samples aside to cool and put an arm around Mary's waist.
"Just let me unhook the water from the stack and then we can go inside and I'll look at your sink. I'll see if I can find that short you were talking about. I need to let everything cool and settle for a while anyway."

Mary nodded. "Good and then I'll fix you some dinner."

Dan went over to the outside spigot and unhooked the hose that had been running the water in to his invention and then followed Mary in to the house. She went straight for the sink and turned on the water. Frowning she turned to look at Dan

"That's funny; I didn't get shocked that time." She stepped away and then back to the sink and tried again. Shaking her head she looked at Dan.
"I swear Dan; I was getting shocked every time I touched that earlier."

Dan shrugged. "If it happens again, just let me know. I'm sure there just has to be a short somewhere."

Mary nodded and got busy fixing dinner while Dan went back out to put everything away.

Over dinner the two talked. Mary always loved their meal times. It was when they had their best conversations. She looked at Dan. "So where do you go with this invention from here? I mean this is a pretty big idea."

Dan nodded. "I've been thinking about that. I'm going to give Joel Benson a call."

Mary looked at Dan confused. "You mean the attorney?"

Dan nodded. "Yeah, I think if this thing does what I hope than I better get it patented. Maybe Joel will have some ideas."

Mary laughed. "Well, it certainly is doing something and you have the video to prove it."

Dan took a sip of his coffee. "The video is just part of it. I'm sure we'll have to have some kind of emissions testing done. I'll ask Joel about that too. I know down at the plant everything we do has to meet certain specifications. If the asphalt we make isn't in spec than it doesn't get used. I'll give Joel a call tomorrow and set up a meeting."

Mary nodded. "Good, that's settled. Now, how about I finish the dishes up and we go out for a movie to celebrate?"

Dan stood up. "It's a deal; I'll even help with the dishes."

*

Two days later Dan sat in the reception area of Joel Benson's office. Joel's secretary, Maggie, looked at Dan. "It should be just a couple more minutes. Would you like a cup of coffee or something?"

Dan shook his head. "No, I'm okay, thanks for the offer though."

Maggie nodded and went back to her typing.

Dan looked around the tastefully decorated room. He noticed a lot of leather bound books in dark, hardwood bookcases. He also noticed most of the furniture was also leather. It looked like Joel was doing pretty well for himself. Dan looked up when the door to Joel's office opened. Joel wheeled himself through the door and smiled at Dan. He then wheeled his chair over to where Dan sat and shook Dan's hand in a firm grip.
"Nice to see you Dan. How are your folks doing?"

Dan smiled back. "Mom's good and Dad's getting old and just as ornery as ever."

Joel laughed. "He was pretty ornery before. Come on in to my office and we'll take a look at your stuff. I have to say you've really peaked my curiosity."

Dan followed Joel in to the office. Dan watched as Joel wheeled his wheelchair behind the large mahogany desk. Dan had deep respect for the man sitting across from him. Joel had lost the use of his legs in a mining accident and after a couple years of feeling sorry for himself, Joel had picked himself up and put himself through law school, wheelchair and all. Dan didn't know what he would have done if he had been put in the same predicament, but felt if he did even half as good as Joel it would still be quite an accomplishment.

Dan had brought a small briefcase to hold all the information on his invention and sat it on the table looking at Joel. "I have the plans here and also a video we shot of the new invention. Do you want the paperwork first or would you like to have a look at the video?"

Joel smiled. "Let's have a look at the video. Then you can try and explain it all to me."

Dan handed Joel the DVD he and Mary had made. Joel took it and put it in a player that sat on a shelf. He opened the large doors above the shelf to reveal a television set inside. The two men sat together and watched. Dan pointed to the TV and explained to Joel what was happening as the video played. Watching the video and seeing Dan opening the door so the fire could be seen, Joel turned to stare at him.

"What the hell did you invent Dan? That looks pretty amazing. Does it take down the gases as well as the smoke?"

Shrugging, Dan shook his head. "I hope it does, but that's one of the things I wanted to talk to you about. I need to get some kind of emission testing done on it. I really don't know where to get that done."

Joel nodded. "There's a few companies around that will do it, but I have to warn you it's damn expensive. Why don't you show me your papers now and explain a little more about this invention to me. Then we can talk about the testing and hopefully the promoting of it."

Pulling the papers out of his briefcase Joel handed the first page to Joel. "That's my incinerator with the stack on top. The stack can be adapted to fit any size incinerator already being used or on a new one. That way you can build them as big or as small as needed." Dan handed Joel another paper.
"That's a breakdown of the stack. As you can see I force the preheated water through a copper coil and then into the water jets. As the water turns to steam it's put through a set of special sprayers to fracture the steam making it small enough to capture the particulates

instead of just pushing them around. The sprayers are crucial as is the way I have them set up."

Joel nodded. "That actually makes a lot of sense to me. You are matching the molecular size so the steam vapor is pretty much absorbing the particulates."
Joel looked at Dan. "If you are taking down the gases too, this is going to be a big deal Dan. Just look at all the pollution going in to the air. Right now the Environmental Protection Agency is fining all the companies that aren't meeting the standards. I'm sure those companies would love something like this to eliminate those fines. Of course stopping the pollution they put in to the air would mean a hell of a lot to people who have to breathe."

Dan frowned. "What about the EPA, what will they think if they lose that money?"

Joel shook his head as he thought about that.
"I'm hoping the benefits to the planet would outweigh the money, but hell Dan, they are the Government and you just never know. I think we'll just take one step at a time. First thing is to get this invention of yours tested and if it is pulling down the pollution and hopefully the gases too, then we need to get you started on a patent."

Dan shook his head. "There's no way I'm going to have the kind of money that'll take."

Joel smiled. "I have an idea, you can think over. I'd like to invest in your invention. I could buy maybe a five percent interest. You can incorporate and make it a non-par company. That means the stock value is zero if your invention flops, but if you make money whatever shares someone owns will go up in value with the value of your company."

Dan frowned. "How do I get money to pay for everything if it's non-par?"

Joel looked at Dan with his intense bluish gray eyes. "You put a value on it. That will be according to what you and your investors feel the invention will be worth in the long run."

Dan shook his head. "Oh hell, I hate to even try and I'd hate to let anyone down."

Joel laughed. "That's one of the reasons people will invest. You're honest and you really do care and probably worry too much because of it."

Shrugging, Dan half smiled. "Mary's always telling me that."

Joel pointed at the papers. "Can I make copies of those? I'd like to check a few things out, talk to a couple of people. Don't worry, I'll ask them to sign a non-disclosure form first, so you want have to worry about someone stealing your idea."

Dan's eyes widened. "Do you really think someone would try that?"

Joel leaned forward in his wheelchair. "Dan if you have what I think you have, there are people who not only would steal that idea, but kill you to see what's in that briefcase."

Dan thought about that shaking his head in wonder and more than a little disgust. Joel was looking intently at him. "This could very well change your life Dan, yours and Mary's. There's a lot of money to be made on an invention that could clean up pollution. That video of yours, burning rubber, do you know how many places are trying to get rid of old tires? That's just one area. The applications here are endless. What would you do with that kind of money?"

Dan smiled and his blue eyes lit up.
"Mary and I would both love to build a place for kids with no home. Also we'd like to have another place for the elderly. I know a lot of older people who don't want to be stuck in a nursing home. Just think how some of the older men would love a fishing pond they could go and sit at. We would both love to have enough money to set up both of those places."

Joel shook his head. "I should have known you would want to do something for someone else. You are one of a kind Dan. I plan on doing all I can to help you. Now let me get Maggie to make some copies of your papers. I do suggest one thing though."

Dan looked at Joel. "What's that?"

Joel shrugged. "I think it would be a good idea if you kept something out of the papers for the patent. It should be something that renders the stack useless without it. Something only you know about. Can you do that?"

Nodding, Dan smiled. "I know just the thing, in fact just give me a minute to redraw one page and it will be done."

Joel nodded. "Good, you finish that and then give me a couple days and let me see what I can come up with."

As soon as Dan left Joel's office he went home and excitedly shared Joel's ideas with Mary. When Dan finished he shook his head. "Wouldn't it be great if Joel is right and this invention takes off? To tell you the truth I hadn't given much thought to the money side of things. But if he's right we could start a kid's ranch and a place for the elderly too."

Mary nodded. "Not to mention cleaning up the air. I don't think people realize what a problem pollution is. What the effects are going to be down the road if we don't get it under control. I think people just figure if they ignore the problem that it will just vanish."

Mary looked at Dan. "So what now?"

Dan shrugged. "Now we wait. Joel's doing some research and is supposed to get back to me in a couple of days. We have to do testing and he thinks I need to get a patent." Dan sighed.

"I think things are going to get crazy in a hurry."

Chapter 4

Senator Buchanan sat behind the over-sized desk drumming his fingers on the polished surface. His eyes were closed. As he slowly opened them a hatred could be seen in their steel color.

Senator Buchanan was a man with his eye on the Presidency. Maybe not in the election coming up, but definitely the one after. A lot of dirty deals were done under his careful watch. Steve Buchanan was always careful to keep his own hands clean and was never linked to the many underhanded plans he laid out. With a determined look on his face, Steve picked up the phone and dialed. He used a private line not linked to his Senate office.

In another office about a fourth the size of the Senator's the phone rang. Betty Mackey wasn't only the secretary; she was the owner's wife. She tossed back her curly red hair as she picked up the phone. "Good morning, Mackey testing, this is Betty. How can I help you?"

The Senator's clipped voice came over the line. "I need to talk to Mr. Mackey."

Betty could tell by the man's voice that he was a person used to getting what he wanted.

Betty didn't have a lot of patience with people like that. "Mr. Mackey is on a project right now, could you hold for a minute?"

She heard a sigh. "I don't have much time and this is important."

Betty rolled her eyes. "Just a minute and I'll see if he's free." Betty pushed a button putting the man on hold and then stood and walked a few feet to a door. She opened it and yelled.
"Jim, there's someone on the phone for you and he says it's important."

Jim nodded and wiped his hands as he walked over to his wife. "I'll get it; maybe it's a testing job. We could use the work." Jim stepped over, picked up the phone and pushed the lit up button.
"This is Jim Mackey."

The Senator stopped his drumming fingers and sat up straighter. "Mr. Mackey, this is Senator Buchanan. I'm sure you know who I am."

Jim raised his eyebrows and ran a hand through his blonde hair. "Sure, what can I do for you Senator?"

Betty stepped closer frowning and watching Jim's face. The Senator was talking. "I've been told you have a testing job tomorrow for a Mr. Warren."

Jim nodded. "I'll be going out first thing in the morning. Mr. Warren has an invention he wants an emission test on."

Steve already knew all that. "How's business Mr. Mackey?"

Jim frowned wondering what this call was all about. "Things are picking up. We're still a fairly new company."

Steve smiled, but it looked more like a grimace than an actual smile. "I happen to know that there are some Government testing contracts coming up and I'd like to be able to put in a good word for your company."

Jim had a shocked look on his face. "I'd really appreciate that Senator."

The Senator's fingers began tapping on his desk again. "There is one thing I need from you though."

Jim closed his brown eyed and shook his head. He knew there's be a catch. "What can I do for you Senator?"

Steve took a breath. "This invention that you will be testing tomorrow, for reasons I can't disclose at this time, cannot be brought forth at this time. I want the actual test results; Dan Warren is not to see them. You need to somehow make the test invalid as far as Mr. Warren is concerned. Believe me Mr. Mackey there is a good reason for this, but like I said, at this time I can't tell you the reason. I promise if you do this for me, for your

Government, your company's name will top the list on any testing contracts coming up. I might add there are quite a few large and lucrative contracts coming up in about a month. I will personally make sure those contracts will be yours if you can see your way clear to do this."

Jim looked at Betty whose green eyes had narrowed suspiciously and then around at his shabby, small office. "Senator, I believe you can call that a done deal. I have an idea that you would like copies of the results am I right?"

Now Steve Buchanan wore an actual smile. "As soon as the paperwork is on my desk, you'll be sent paper work on the testing contracts. I believe this is a win, win situation. It will be of benefit to both of us Mr. Mackey."

Jim wasn't sure, but he needed the business and nothing paid better than Government contracts. "Thank you Senator, I'll have the results in a week."

The Senator nodded, he was used to getting his way. "One more thing, I think it will be better if we keep this to ourselves."

Jim was looking at Betty when he answered. "I think so too."

Jim hung up the phone. Betty stood looking up at him, her arms folded. "What's going on Jim?"

Jim tried to smile. "Actually it's good news Betty."

Betty looked at Jim's face and knew he wasn't telling her something. She also knew it was better not to press him. Sooner or later he'd tell her. Jim wasn't a secretive person.

Betty smiled."Okay, so tell me the good news. Was that really a Senator?"

Jim let out a breath as he thought what all he should share with his wife, for now at least. Jim gave Betty his best smile. "Yeah, it was Senator Buchanan and he offered to line me up with some Government contracts. That could be just what we need to get the company up and running."

Betty frowned. "What's the catch?"

Jim shrugged. "No catch, the Senator just wants copies of my reports. Probably just wants to check out the quality of my work." Jim put an arm around Betty's shoulders. "Listen, I promise there are no secrets about all of this. Someone had to of told the Senator that I do good work and probably cheaper. Don't lift a gift horse in the mouth honey."

Betty shrugged. "Maybe you're right. I just didn't like the Senator's tone."

Jim laughed. "He's probably just used to getting his own way. You know how those guys are. They think everyone is in awe of them and expect everyone to bow down at their feet."

Betty shook her head. "Not me, I'm not impressed by politicians. Personally they're all a bunch of blow hards to me. I don't trust any of them."

Jim nodded. "I feel the same way, but we really could use the business."

Betty shrugged. "What about this job tomorrow? Is it still on?"

Jim nodded. "Yeah, actually I'm quite excited to see this Warren guy's invention. It's some kind of new pollution abatement device. Most of the ones I've seen don't do a hell of a lot. I think I'll take Benny with me. If we do get some Government contracts I'll need him full time."

As Betty was getting ready to ask Jim more about that, the phone began ringing. Jim smiled. "You better get that, I have to go get the equipment ready. Just give me a holler if you need me."

Jim walked off as Betty picked up the phone.

The next morning Mary was pouring Dan's coffee. "What do you want for breakfast?"

Dan shook his head. "No breakfast, I really need to go and get the stack ready. I want to heat the water before Jim Mackey gets here."

Mary looked at him frowning. "You should still have breakfast."

Dan held up his cup. "This will do for now. After the testing is done, we can go out to lunch somewhere."

Mary nodded. "It's a deal. Can I come out and watch?"

Dan stood up and drank the last of his coffee. "I'd be put out if you didn't."

Mary smiled. "I'll be out in a minute. I just want to straighten up the house."

Dan leaned down and kissed her cheek. "I don't think there's much straightening to do, but I'll be outside when you're ready."

Mary smiled as Dan walked out the front door. She walked in to the living room thinking that maybe Dan didn't see the need for straightening, but she wasn't going to take the chance of someone stepping in to her house and thinking it wasn't clean. After finishing the living room, Mary went to the bathroom. If Jim Mackey needed to borrow the bathroom Mary didn't want a mess. She never could stand a messy room and God forbid someone came in and saw one besides that the cleaning kept her busy. She was almost certain she was more nervous than Dan about all this testing stuff. She knew how much it meant to him. First off it would be a validation for Dan that his invention worked and then the possibility that something he had invented could not only change the world, but make enough money to actually start a kids ranch and a place for the elderly was a dream come true. Mary and Dan had been having that dream for years. They had always figured that's what their idea was, just a dream. Something like that was just beyond regular working folks. Mary shook her head and returned her thoughts to her cleaning.

Going to the bathroom sink, Mary was surprised to get shocked when she reached to turn on the faucet. She frowned. This hadn't happened since the day she had told Dan about it when he had been first running the stack. Mary sighed; it would just have to wait until later. She looked around; satisfied that things were in order and went outside to join her husband.

An hour later a full ton black Chevy truck pulled into Dan and Mary's yard. It had a utility bed and was pulling a silver trailer behind it. On the side of the door blue letters spelled out 'Mackey Testing' along with the company's phone number. Dan and Mary watched as a blonde man about thirty got out of the truck, followed by a younger man who, if judged by looks, had to be Jim's brother. The two walked over to where Dan and Mary were standing next to Dan's invention.
The taller of the two men put out his hand toward Dan. "Hi, I'm Jim Mackey and this is my little brother Benny."

Dan shook first Jim then Benny's hand.
"I'm Dan Warren and this is my wife Mary."

Mary smiled. "It's nice to meet both of you."

Jim looked up at the tall stack. "So, I'm guessing this is your invention."

Dan nodded. "This is it. I've just about got it ready for us to start the burning."

Looking at Dan, Jim nodded. "What are you burning in there?"

Dan shrugged. "I thought I'd burn some coal at first and then I'd like to throw in some rubber belting to simulate tires."

Jim nodded as he turned to Benny. "We better go and get things set up."
Jim turned back to Dan. "We should be ready in fifteen minutes or so."

Dan smiled and then sighed. "I'll be ready I guess. I have to admit I am a little nervous."

Jim could see the worry in Dan's face and could feel his own guilt rise. That was one thing he couldn't allow to happen. He needed to think about his company and the Government contracts that would be coming his way. He tried to smile. "No need to be nervous, let's just see what happens."

Jim and Benny walked away from Dan and Mary and over to their trailer.

Dan got busy building the fire in the incinerator. As soon as the coal started burning, black smoke came billowing out of the stack's exhaust on top. Dan looked at Mary who was frowning. He laughed. "Don't worry; the invention will kick in soon."

Dan reached over on the table next to them and grabbed the video camera and handed it to Mary. "Here, you can be the official camera person. Make sure to get a video of that dark smoke before it's gone."

Mary nodded, if she hadn't seen the invention before she would have thought Dan's words were crazy. Mary walked back where she had a good view of the stack and the best camera angle, and then she began filming.

Jim and Benny came out of the trailer carrying their equipment. Both men looked up at the stack as the black smoke began turning to gray. They walked over to Dan. Jim pointed up. "Does it get lighter than that?"

Dan nodded. "Just watch, it should just take a couple more minutes."

Jim and Benny set up ladders on either side of the stack. Benny pointed up at the top of the stack. "Oh my hell, would you look at that?"

Jim looked up and shook his head. The smoke had vanished. He turned to Dan. "Is your fire still going?"

Dan laughed as he bent down and opened the door to the incinerator. "What do you think?"

Both Jim and Benny's eyes widened in surprise at the large flames in the firebox.

They could see the coal embers burning bright red inside. Jim turned to face Dan.

"Looks like your invention is at the very least taking down the particulates. Now Benny and I will go ahead and start the emissions testing."

Dan nodded. "Do you need me to do anything?"

Jim shook his head. "Not yet, but I will ask you to periodically open that door for me. After we do the first testing, I'll have you throw in some of that rubber you were talking about and we can do some testing on that."

Benny and Jim grabbed their long probes and took them up the ladders to the top of the stack.

About a half an hour later a vehicle pulled in to the driveway. Dan smiled as he recognized Joel's van. A man got out from the passenger side and walked around to where Joel sat in the driver's seat. Dan saw the man nod a couple of times to Joel before turning and walking toward him. The man put out his hand as he stepped up. "Hi, you must be Dan Warren. I'm Ron Letner, I run the State Economic Development Center. Joel has only good things to say about you and your invention here. I hope you don't mind that Joel brought me out to see it."

Shaking the man's hand, Dan shook his head. I don't mind at all." He turned to look at Joel, who was lowering the lift on the side of his van.

"Does Joel need any help?"

Ron shook his head. "I already asked and he said he has it down to a science."

Dan nodded. "Okay then. Come on over and I'll show you what we're doing."

Dan looked up to where Jim and Benny were busy with their testing. "Would it be okay if I opened the door to show Ron the fire?"

Jim looked down from his place on the ladder. "Just let me finish up this one test and then I'll need you to open the door anyway."

Jim reached over and turned a couple of knobs and adjusted his probes, and then he looked at Dan. "Okay Dan, you can go ahead and open it now."

Reaching down, Dan opened the incinerator's heavy door. Ron shook his head in amazement as he gazed in at the fire and then up to the top of the stack. "This is amazing Dan. I can't believe there's no smoke coming out of there."

Dan smiled, he was pretty amazed himself. He never thought the invention would work this well. "I'm going to go ahead and throw in some rubber belting next. I really have to tell you this stack would be great sitting on top of some of the big time incinerators being used now."

Ron was nodding enthusiastically. "What they have in use now is a big problem alright. There's too much pollution being emitted. If you can eliminate that or even suppress most of it. You will really have something."
He looked at Dan. "I would really like you to come into the Economic Development Office. I think we can get you some grant money on this invention." Ron reached for his wallet and pulled it out and grabbed a business card out of it. "Just give me a call and we can set something up." Ron looked up to the top of the stack again and shook his head. "This invention of yours is going to be big. I can already think of quite a few different applications to try it on."

Dan nodded. "I've been thinking about that myself."

Both men turned as Joel wheeled up. Dan smiled. "Good to see you Joel."

Joel smiled back. "Looks like you've met Ron. If anyone can help you get your invention going it's Ron. Also I talked to a patent attorney who can help you get your invention patented."

Dan raised his eyebrows as he shook his head. "I can't believe it; I don't know how to thank you for all your help."

41

Joel shook his head. "My part was easy; you did all the hard work. Just remember, there's a hell of a lot more to come. Between patenting your invention and then promoting it, you'll feel like you have a full time job with it."

Dan laughed. "I already have one of those."

The men looked up as Jim shouted down at Dan. "If you're ready, go ahead and throw that rubber in."

Dan went over and grabbed some of the rubber belting and began adding it to the fire. All eyes looked to the outlet of the stack, but no smoke came out.

On top of the ladder, Benny turned to Jim.
"I can't believe this. No emissions are registering." He looked at the gauges. "I see a slight amount of carbon monoxide and sulfur, but what this invention is doing is groundbreaking."

Jim shook his head. "Maybe there's something wrong with the probes. Nothing works that well. I'll have to wait until I get back to the shop before I know for sure, but I think this test may be a bust."

Looking at his brother, Benny frowned. "How can you say that? You can see with your own eyes that nothing is coming out."

Jim wasn't sure what to say, but he knew the Senator was going to be very interested in the findings and definitely wouldn't want Dan Warren to know about them. Jim just shrugged.

"I'll wait and see what the computer readings are before I have any conclusion. I think it would be better if both of us waited. I wouldn't want to tell Dan Warren one thing and then have the tests come out something different."

Benny glared at his brother. Something was wrong and he was pretty sure it wasn't the equipment messing up. Benny knew he had better just drop the subject though. If Jim wanted to share whatever was going on Benny knew he'd do it in his own good time and he would just have to wait him out.

*

Three hours later, Dan and Mary were sitting down to a chili dinner. Mary had started the crock-pot that morning and now everything had cooked and blended just right. She looked at Dan across the table. "Well, what do you think? It went pretty well didn't it?"

Dan nodded. I think so, although it took a lot longer than I thought. I was going to treat you to lunch."

Mary shook her head. "I'd rather have this chili anytime."

Dan smiled. "Me too, it is delicious by the way." He took a bite and then smiled at Mary. "I think one of the best things about today was meeting Ron Letner. He was really impressed. I think he'll get me some grant money to help with expenses. God knows I'm going to need it, the testing for today alone is over ten thousand dollars."

Mary dropped her spoon and her eyes widened. "Testers make that kind of money?"

Dan nodded. "Sure, but you gotta figure Jim Mackey probably has a fortune tied up in that testing equipment of his. Plus it's all linked to computers, so I guess that even charging that much he doesn't make much of a profit."

Mary smiled. "You think the best of everyone Dan. You could have thought you were being overcharged, most people would have."

Mary blew her husband a kiss.
"You are a nice guy Dan Warren."
Then she shook her finger at him. "Just remember a lot of people will take advantage of that fact. You know the saying..."

Dan laughed. "I think you may have mentioned it once or twice. Nice guys finish last. I know and I'll be careful."

Mary smiled. "Good, now finish your dinner." Then Mary frowned. "Oh Dan, I forgot to tell you I got shocked again this morning. This time it was in the bathroom at the sink in there. It's the weirdest thing because last time it happened you were working on the stack too. Remember you were doing the trial run."

Dan looked at Mary and she could see wheels turning. "I remember and it is weird. I might have an idea about why though. Let me think on it and I'll let you know.

Jim Mackey was alone in his shop about the time Dan was thinking about what Mary had told him. Jim had sent Benny home right after they had finished the job at the Warren's and had just told Betty to go ahead and leave early and he'd shut the shop up. Right now, Jim just wanted to be alone. He hated himself for what he knew he had to do. He knew that he was going to have to tell Dan Warren that the test was invalid. Jim wasn't much of a liar and doing all of this wasn't sitting right with him. He only hoped that by the time another set of tests could be done on that invention the Senator would have changed his mind about the whole deal. Jim could see how important an invention like this could be. There had to be a good reason for the Senator asking Jim to not reveal the actual test results to Dan. Jim hoped so because as he looked at the computer print outs he was amazed. The invention had only let a tiny percent of gases slip in to the air and an even smaller amount of the particulates. Jim shook his head in wonder at what was going on inside Dan Warren's invention.

How the hell was Dan doing it? Jim shrugged; it wasn't his business to know how, just to get the results.

Jim made an extra copy of his findings. He put the copy in an envelope and locked it in his safe, sliding it under some papers already in there. He felt guilty even doing that, he hated keeping secrets.

The original results he put in a large envelope and addressed it to the Senator.

Tomorrow he'd send it out with a delivery confirmation attached to make sure it went straight to the Senator.

Jim shook his head wearily. He hoped he was doing the right thing. Jim turned off the lights, locked the door and headed home.

Chapter 5

Dan sat in the front room of the State Economic Development Office. He held a magazine, but had already read the first line of the article at least fifty times and couldn't remember one word of the text. He was too nervous to concentrate. Dan looked up when he heard someone calling his name.

Ron Letner stood in his office doorway waving at Dan. "C'mon in Dan and let's see what's going on or what we can get going on."

Dan walked over and shook Ron's hand.
"Thanks a lot for seeing me."

Ron laughed. "Don't thank me; you're the one with this great invention. I should be thanking you."

The two men walked together into Ron's cramped office. Dan looked around the room at the file cabinets stacked along the walls.
There didn't seem to be a space that wasn't covered in paper stuffed boxes and files. Even the drawers on the file cabinets had papers sticking out of them to the point they wouldn't shut.

Ron pointed at the only chair besides his own in the office. "Have a seat Dan."

Dan sat and watched Ron as he made his way around several boxes to get to his chair.

Finally he took his seat and Dan studied the man a moment. With his white hair and receding hairline, Dan figured Ron must be in his late fifties. Although looking at Ron's round face, Dan saw very few wrinkles around his eyes that were a light gray and were animated with excitement.

Ron was smiling at Dan. "This invention of yours is going to be big Dan. I've had a lot of different inventions and ideas come through my office."
He pointed at the overstocked file cabinets.

"I have to tell you, nothing I've seen comes even close to what you have. The potential is enormous.
Ron pulled a folder out of his desk drawer.
"I've been doing some checking and it looks like there are quite a few different applications your stack can be used for. Garbage, tires, coal and a lot more. We have grants for the new industrial development and manufacturing. The manufacturing could be used later if you decide to build the invention and sell it to companies to use."

Dan shrugged. "Right now I need to pay for a patent and more testing."

Ron nodded. "I still think I can get you a fifty thousand dollar grant for development. I know that isn't really that much when you probably have some big bills for the testing you've had done and need to still have done. We can start with that though. I think I can have something two weeks after I submit the papers. I have to admit I have already filled out a form. If you want to look it over and if it all looks okay, sign it. I think we can get it started right now."

Dan opened his eyes wide. "Just like that?"

Nodding, Ron laughed. "I've seen that thing of yours working remember. I believe in it. When you get the results of that test back I can probably swing another grant. Believe me Dan; this invention of yours is going to be the biggest thing to hit this country."

*

An hour later, Dan was at home trying to explain everything that Ron had said to Mary. "This whole thing is just crazy Mary. I never even thought much beyond getting the proto-type built. Now Ron has me right up there with Tesla or Einstein."

Mary half laughed. "Just remember a lot of Tesla's works got stolen or suppressed. I'm sure even now his inventions are hidden in a dark Government room somewhere."

Dan shook his head. "Well the way Ron talked, that will never happen to my invention."

*

Ron was still in his office. He worked late a lot. Ever
since his wife had died five years ago, the job had
become his life. Sometimes he wished they had been
able to have children. That just hadn't been in the cards
for them and they had been happy to just have each
other. Maybe that's why they had been so close and why
he missed her so much. His wife had gotten cancer and
Ron had watched as his beautiful, vibrant wife had
shriveled to a fourth of her former self as she succumbed
to the ravishes of cancer. Then his better half had died,
leaving him alone. It had been the hardest thing Ron had
ever faced and to this day he thought about her every
morning as he woke and turned to look at the empty side
of the bed. Ron sighed and shook the thoughts from his
head. Focusing instead on the invention Dan Warren had
brought him. Ron had never come across anything like it.
He was sure Dan Warren didn't realize what he had
either. Dan had focused on a way to burn landfill
garbage, which was great. If every landfill in the country
had an incinerator with Dan Warren's stack on it, the
invention was worth millions. If you added in tires and
coal incineration with no pollution, the possibilities were
endless. If Ron could talk Dan into manufacturing the
stacks in this state a lot of jobs could be created. Ron

took notes as the various scenarios ran through his head. By the time he got ready to call it a night he had already filled quite a few sheets of paper with his ideas.

*

The next morning, Ron was back at his desk looking once again at the notes he'd written down the night before when his secretary buzzed him. "Ron, Senator Buchanan is on line one."

Ron pushed the intercom button. "Thanks, I got it." He hit line one and picked up the phone.
"Ron Letner here."

The Senator's staunch voice came over the line. "It's Senator Buchanan; a little problem at your office has come to my attention."

Ron frowned. "What kind of problem?"
Ron didn't like Senator Buchanan but knew he held a lot of strings in his slimy fingers.

"I'm told you are trying to get a grant on a Pollution Abatement Device for Dan Warren. I have talked to several important people in Washington and right now I'm afraid that is going to be an impossibility."

Ron's mouth dropped open. "What are you talking about? I already have the paperwork ready to send off and I have talked to Dan. I pretty much guaranteed him a grant."

The Senator's word came back low and slow and with a bite to each syllable. "Then you need to un-guarantee him."

Ron heard the Senator take a deep breath over the line. "Listen Ron, this comes from high-up. There will be no invention at this time and no grants. I am not at liberty to go in to any details, but I have to say if you try and go ahead with getting Dan Warren a grant you could very well be putting your job in jeopardy."

Ron was in shock at the Senators words. He also knew he wouldn't be getting an explanation. The Senator wasn't a person who felt he had to explain himself to others. Not only that, but he knew he would be the one to have to break the bad news to Dan. "Okay Senator, have it your way. For the record, I think you are doing the wrong thing."

The Senator was smiling, but his words were still cold. "Think what you want, but remember the invention is to go no further."

The Senator hung up and Ron stared at the phone a minute before placing it back on its base. He stared at his notes on the invention and shook his head wondering how in the hell Senator Buchanan even knew about Dan Warren's invention or the fact that Ron wanted to get him a grant. He had only found out about the invention through Joel. From what Joel had told him only a few select people were being told about the invention right now.

Ron sighed, now it was up to him to tell Dan the bad news. Ron stared at the phone in indecision. He knew he should just get it over with, but this was one call he wasn't looking forward to. He hoped the Senator had a damn good reason for suppressing an invention that could be such a benefit to the world.

Ron picked up the phone as he decided it was better to just get it over with. Dan answered on the first ring. "Hello."

"Hi Dan, it's Ron."

Dan smiled. "Hi Ron, you just caught me. I was just headed out to work."

Hearing that, Ron wished he would have waited a few more minutes. "This won't take long Dan. I'm afraid I have some bad news. I am really sorry about this Dan, but I am not going to be able to get you that grant I thought I could. Government red tape is about what it boils down to. But we can't put out the money on just an invention. If you were already a business we might be able to put some funding together, but as it is I just can't find a way to get you a grant. I can't tell you how bad I feel about this. I still feel you are sitting on a gold mine with that invention of yours. I also think someday it will change the way we operate. Please Dan, whatever you do, don't let anyone stop you from bringing this invention to public awareness."

Dan had to laugh. "You're worrying too much Ron. I'll figure something out. I think Joel is checking out a few more avenues I can try."

Sighing, Ron nodded. "Joel does know a lot of important people. I think he'll be able to come up with something. I'd still love it if you would keep in touch and let me know what is happening though."

Dan nodded. "I'll do that Ron."

Dan hung up the phone just as Mary stepped up. "Was that Ron Letner?"

Dan nodded. "Yeah, it looks like we're a no go on that grant business for now. I guess if I ever make a business out of building the stacks or something he might be able to help then."

Mary shook her head. "You start manufacturing those stacks and you won't need a grant."

Dan smiled. "You're probably right. We just have to figure a way to get from inventing to building."

Dan bent down and kissed his wife. "Right now though, I need to get to work. I'll see you later."

*

Mary and Dan were seated at the kitchen table that night when the phone rang.
Dan stood up. "I'll get it; maybe it's Joel with some good news."

Mary nodded as Dan went to get the phone. "Hello."

Jim Mackey's voice came over the line. "Hi Dan, it's Jim Mackey. I'm afraid I have some bad news for you."

Letting out a breath, Dan rolled his eyes. This was definitely not one of his better days. "What's going on Jim? Is it something to do with the test? I thought it went pretty good."

Jim felt like kicking himself for what he was about to say, but felt he didn't have much of a choice.
"I think your invention performed perfect Dan. It was my machines. The testing didn't turn out. I'm thinking it had to be some kind of computer glitch. I'm having someone come in and check out the machines for me."

Dan felt like someone had kicked him in the stomach.
First the call from Ron and now this.
He felt like some invisible force was trying to stop him from getting his invention going.
"Will you be able to retest? What about the charge on this test? I know it didn't work, but I'm sure you still have to charge for your time and the use of the testing

equipment. Am I paying for that? I mean I know I'll have to pay, but will the charge be less. I hate to even ask, but I'm in way over my head here."

Jim shook his head. "No, I can understand where you're coming from. The charge will be as low as I can get it. This is my fault and you shouldn't have to pay for that. When I get my computers fixed we can try and set up another test."

Dan nodded. "Thanks for letting me know Jim. Could you give me a call when you're ready to do testing again?"

Jim nodded. "You bet Dan."
Jim hung up the phone feeling like the biggest heal of all time.

As soon as Dan stepped back into the kitchen Mary could see something was wrong. "What's going on hon, who was on the phone?"

Pulling out his chair, Dan sat down shaking his head. "That was Jim Mackey. The testing somehow got screwed up."

Mary frowned. "What does that mean?'

Dan shook his head again. "To tell you the truth, I'm not sure. I overheard Jim's brother telling him that the emissions coming out were pretty much non-existent. According to Jim though, the computer readings got messed up and the test didn't register. Jim said it had to be a glitch in the system."

Mary looked at Dan is disbelief. "Something wasn't right, but I don't believe that it was the computer. If you add in this deal with Ron Letner falling flat it all sounds a little fishy to me."

Dan smiled. "I think you've been watching too many conspiracy shows."

Shaking her head, Dan could see the anger in his wife's usual bright blue eyes. "That's not what it is Dan." Mary took a deep breath. "You need to get another test done."

Dan nodded. "Jim said he's having someone look at his computer."

Mary shook her head. "No Dan, I think you need to find another company. Maybe Jim's computer was bad, but they always say you should get a second opinion. In this case I think you should."

Dan was quiet a moment and then he nodded. "Maybe you're right."

Mary laughed. "Of course I am."

*

A week later, Dan was sitting across from Joel in his office. Joel was shaking his head. "I'm really disappointed with Jim Mackey and Ron too. Of course I don't really blame Ron. His job is one big headache after another. He has to try to go along with all the rules and regulations the state sets down and the Federal governments is right there monitoring everything too. If there would have been a way to get you some money I know Ron would have found it."

Dan shrugged. "It's okay, I understand. I'd still like to try and set up some more testing though."

Joel smiled. "I've got an idea on that. There's a physicist at the college. His name is Trent Altman. He has a radio show, maybe you've heard it."

Dan frowned and shook his head.

Joel continued. "He's a brilliant scientist, although a little crazy. Some of his theories are pretty far out there. I believe in most of them though, I have to admit. Anyway, I really think you two should meet.

Trent may be able to help you out quite a bit. He can line up the testing for your stack through the college. They can use it as a project. That way the charge should be minimal. You would have to pay for any extra equipment they might need, maybe you could even rent it instead. I can't think of much else they would need."

Dan's face lit up. "That sounds great. I won't have any free time for a couple of days, but after that, I would be glad to meet with this physicist."

Joel nodded. "I'll get something set up. I'd really like to be there. I think a conversation between the two of you is one I don't want to miss."

Dan shrugged. "I don't know about that, but I'd be more than happy to have you there."

Joel smiled. "Tell you what, why don't we all meet here? I already have copies of your paperwork and the DVD of you running your stack. We can go over everything with Trent and see what he thinks. Also I talked to a patent attorney and he can start on your patent right away."

Frowning, Dan shook his head. "Even without a legitimate test? Not to mention some money."

Shaking his head, Joel grinned. "I paid the retainer and don't worry, I'll trade that for stock. You don't need the test anyway because you're patenting the idea. The lawyer will be calling you, but everything is already in motion."

With a sigh, Dan shook his head. "I don't know how I'll ever be able to get even with you."

Joel laughed. "Don't worry about that, I believe in your invention. But more important, I believe in you.

*

That night after dinner, Dan turned on the radio. Mary walked over as he scanned the channels. "What are you looking for Dan?"

Dan turned to her. "Joel was telling me about this guy, Trent Altman. He's a physicist over at the college. He also does a radio show. Joel is having me meet with him in a couple of days. I thought I'd listen to his show, kinda get a heads up on what he's all about."

Mary reached over for the tuner on the radio.
"Here, let me do that, I know where the college radio station is."

She played with the radio for a few seconds. The voice that came over the radio was deep and rich and Mary pictured a big man behind it.

"...like I was saying, a lot of tests and experimenting have been done on water molecules. You need to know that we have only just scratched the surface. Right now I am starting a company that is adding an electrical charge to ordinary drinking water. I have to admit we are having fantastic results in curing some common ailments, like arthritis and even weight loss. Remember when you were told to give kids Pedialyte when they were sick? It is the same principal here. I know of another company that is testing ice crystals when they are exposed to different natural elements. Even the sound of various types of music is showing a difference in the make-up of the molecular structure. Under a microscope, the crystals change with each new exposure. Nature has given us a vast supply of water and it is up to us to find applications for its use. Even Sea water has a significantly higher electrical charge that can be used to our advantage without tapping in to the water supply we need for drinking."

Trent Altman went on with his show, but Dan had stopped listening. His mind had caught on Trent's comments about water and electricity. His mind went back to Mary getting shocked every time he had run the stack. He had used the outside garden hose for his supply of water. Now he wondered if somehow that had been what had caused the static electricity in the house. Dan's mind was whirling. If the stack was somehow making electrical energy through its process, then the invention was doing a lot more than he ever had expected. Dan saw Mary looking at him curiously. He smiled back and was going to try and explain when both of them heard the loud rumble of thunder.

Mary looked toward the kitchen window.
"Guess that thunderstorm they've been predicting is finally going to hit."

Dan walked over to the window and looked out as a streak of lightning filled the sky.

Dan watched the jagged bright light as it streaked through the dark sky to the ground. It was easy to follow the trail of the bright light as it made its crooked presence known.

Dan smiled, thinking that lightning wasn't too different from his invention. The water he used in steam form was causing a static electricity that was helping to attract the particulates and gases in the smoke and tear them away from the smoke from whatever he happened to be burning. Lightning too had energy, powerful energy.

What in the hell had he invented.

Chapter 6

Dan and Joel sat in Joel's office waiting for Trent Altman. It was after hours, so they were the only ones in Joel's office. Dan had just finished telling Joel the same story he had shared with Mary a few nights earlier.

Staring at Dan, Joel shook his head.
"That's incredible Dan. If you are making energy in that stack and bringing down pollution in the process, we got a whole new ball game going on."

Dan nodded. "I think that's why it is working as well as it is. However I'm making that charge it is like static cling and the particulates are being drawn to it. I think it is capturing twice as much as the steam alone. I'm hoping that Trent Altman with his background will be able to fill in some gaps for me."

Joel looked at his watch. "He should be here any minute. I'm going to set up that video on your stack. I'd like him to watch that first."

Dan nodded as Joel wheeled his chair over to the cabinet that held his TV.
Dan went through his paperwork, putting it in order.

Before he finished there was a knock on the door.

Joel smiled and yelled. "C'mon in, the door's open."

When the door opened Dan looked over at the man who stepped through. He was probably in his early thirties with a bald head. Dan couldn't tell if nature had given it to him or if it was self-inflicted. The man had light brown eyes behind thick glasses that made them look extremely large. As Dan stood he saw Trent was about two inches taller than his own six foot two. He was also about forty pounds lighter. Dan held out a hand and stepped over to Trent. "You must be Trent, I'm Dan."

Trent smiled. "You are the man I am here to see. Joel has some great things to say about you."

Dan laughed. "What's he's told me about you is pretty interesting too."

Joel looked at the two men. He was thinking the next couple of hours of conversation was going to be enlightening and hopefully life changing for a lot of people. "Why don't you both sit down and we can watch Dan's video and then we can go from there."

Both men took chairs as Joel started the movie. Dan explained the Pollution Abatement Process as the movie played, but left out the electricity he felt the stack was creating. He wanted to hear what Trent had to say first.

When the video was over, Trent was silent a moment, then he turned to look at Dan. "Have you had your water checked for an electrical charge yet?"

Dan's eyes opened wide. "No, it hadn't occurred to me until the other day that there even might be one. In fact, it was your radio show that got me thinking about it in the first place."

Trent nodded. "What do you contain your water in?"

Dan shrugged. "Right now I am using large plastic containers to hold it in. The smaller samples are in glass bottles though."

Trent continued to stare at Dan. "I want samples, but I want them from the plastic containers. You can transfer them in to smaller heavy plastic bottles if you need to. I have some you can use. I've found that they hold the electrical charge better. If you have one it is easy to find out. I, for one, would bet that you have a charge."

Dan nodded. "I think I do too, but you know a lot more about it than I do."

Trent laughed. "I have a PHD, but there's no way I could have built an invention like yours. Don't sell yourself short. There is more than one kind of knowledge Dan. Having a few letters behind your name doesn't necessarily mean you know more. I have to say it does help open doors though."

Dan nodded. "I didn't even go to college. I've been working since before I got out of high school."

Nodding, Trent smiled. "That's what I'm saying Dan. You've been out in the real world. You see that something needs fixed and you take care of the problem. To me it looks like you must be a damn good fixer."

Shrugging, Dan shook his head. "I fixed this stack alright. I fixed it so good that I don't know what the hell I'm doing inside of there."

Trent smiled. "I'm hoping I can help with that. First, let's talk basics. You have the water, two parts hydrogen and one part oxygen. Now you can add a hydrogen molecule and get heavy water. That's not always stable. There's been a lot of research on hydrogen. I think that you're dealing with the oxygen molecule though. Oxygen is a perfect balance of electrons and protons, eight is the magic number. Now if in your process you are creating an imbalance, then those atoms are going to start spinning. If the imbalance isn't corrected they will just keep going and your by-product is energy in the form of electricity."

Dan frowned. "Wouldn't that be perpetual energy?"

Trent smiled. "Exactly, but nature is going to try to correct the imbalance. If it does, then your motion stops and with that, the energy dies and it and your water become neutral. I've found the heavy plastic containers are able to retain the electrical charge best."

Dan laughed. "That's funny, because when I decided to use the large plastic containers it was more for convenience than anything else."

Trent laughed also. "Coincidence can be a wonderful thing."

The two men continued talking and Joel sat back in his wheelchair, just enjoying the show.

Chapter 7

Mary stopped talking and Chad reached over and turned off the recorder. "I'm sorry Mary for letting you go on so long. You're probably tired."

Mary rubbed her neck. "Maybe a little, but it really feels good to get it out. At least that's what the therapist they make me see tells me."
She laughed. "Actually, I don't tell him a damn thing." Mary looked slyly at Chad. "To tell you the truth, I think he's crazier than they say I am."
Then Mary laughed and her blue eyes lit up and her face looked years younger.

Chad laughed with her, marveling at the change. What had these people done to this woman? Chad stopped laughing and frowned. "How did you know about the Senator?"

Mary too had stopped laughing. Now her face looked weary again. Chad already missed the Mary he had just been given a glimpse of. Mary shook her head, and brushed back her shoulder length brown hair. "I don't have any proof, nothing on paper or a video, but most of the information came from Trent. At one time he was a permanent fixture in a lot of Senate committee meetings."

Chad frowned. "Was?"

Mary nodded. "Yeah, Trent's story is a pretty interesting one too. He ended up being black balled in the scientific world." She looked at Chad and he could see the genuine sorrow in her eyes.

"It's a damn shame too, Trent is a genius. The Government didn't like his ideas any better than they liked Dan's."

Chad sighed. "I'm sorry for all you've been through Mary." Chad reached on the table for his recorder. "I better pack up and call it a day. If I'm here too long they won't let me back."

Mary gave him a sad smile. "You will be back won't you?"

Chad nodded. "You can count on it. I'll be back tomorrow morning. I want to go and get all you've told me typed up and then show it to my editor. I'm thinking when he sees it; old J.C. is going to want more. He'll probably want to do more than one article. Hell, I'm sure your story would make a hell of a novel."

Chad put all the things he had brought in with him back into the pack he had been carrying and stood up. "Thanks Mary, I don't know when I've had a more interesting time."

Mary looked at Chad. "You know what? To be honest, I should thank you. I can't tell you how amazing it is to have someone want to listen to my story. I guess that's not quite true. I've had people want to listen, but for all the wrong reasons."

69

Mary smiled and Chad saw that special person who was the real Mary Warren shining through. He smiled back. "It's been my pleasure. Now, if I don't get out of here, we'll both be in trouble. I'll see you in the morning."

Mary nodded and Chad left her alone in the room. For Mary the room felt twice as empty as it had before and for some reason cold. It was like Chad had taken the warmth with him when he left. Mary sighed; it was going to be a long night.

Chapter 8

Chad went straight to the newspaper office. He knew he should have gone home to get everything typed up first, but he just couldn't wait to share Mary and Dan's story with J.C. When Chad walked into the newspaper office, he saw four people busy typing, which for 'The Vicinage Voice' that was pretty much a full crew.

Chad looked across the room to where J.C. Monroe sat in his glassed in office. Chad could see J.C. was on the phone. When Chad stepped closer, J.C. saw him and motioned for him to come in to the office. When Chad walked in J.C. held up a finger, rolled his eyes and shook his head. Chad assumed the phone call probably wasn't too important. When J.C. replaced the phone receiver, his first comment confirmed Chad's thoughts.

"Mrs. Carlyle again. I wish even one of her stories were true. I'd be a millionaire selling them."

Chad laughed. "What was it this time?"

J.C. took a breath and wearily shook his head. "She says little men have been breaking in to her food boxes. They only tear one corner and she can see the crumb trails they leave behind when they steal the food."

Chad laughed again and J.C. held up his hand.

"Wait, there's more. Mrs. Carlyle can hear these little men talking, but they are so small she says it only be heard as a squeak to human ears."

Now Chad burst out laughing, leaning forward with the effort. "Did she happen to mention the little men also have four feet and a tail?"

J.C. joined in the laughter. "She's never actually seen one; apparently they are good at hiding."

J.C. shook his head. "It's really kinda sad when you think about it. She's just a lonely old lady. She drives me nuts with her calls, but maybe I'm the only one she has to talk to."

Chad nodded. "I never really thought about it, but I think maybe you're probably right." Chad smiled. "But after what Mrs. Carlyle shared maybe what I have to say won't seem so far-fetched."

Chad was still standing. J.C. pointed at a chair. "Well sit down and let's hear about it. But first, what's she like? I have heard some crazy stories. I'm sure most of them are just that, crazy and a cover-up for what went on."

Chad nodded and was silent a minute as Mary's face came to mind.

Finally he smiled at J.C.
"She's in her thirties. But when she smiles she looks twenty. I have a feeling she doesn't get a chance to do that much. I'd have to describe her as more cute than pretty. She has one of those innocent, baby faces. You can see that haunted look in her face though, especially when she talks about her husband, Dan."

Chad looked at J.C. and leaned forward in his chair, his face serious. "I believe her J.C., every damn word and we've only started. She says the government killed her husband because of his invention and I really think they did. She's locked up in that psychiatric hospital, but she's not crazy."

J.C. ran a hand through his white hair. "Did she happen to tell you she pulled a gun on Senator Buchanan?"

Chad's eyes widened as he shook his head.
"No and I never read anything about that in the reports either."

J.C. shrugged. "From what I'm told, the Senator didn't press charges. They put her in that place because she was a danger to herself and others."

Chad shook his head. "I don't buy it J.C., from what I've heard so far this invention that Dan Warren had would have meant cheap, if not free energy. We both know the government with their oil industry backing wouldn't want that."

Chad let out a sigh. "I recorded her and I want you to listen to that tape J.C., Mary is telling the truth."

J.C. gave Chad a crooked smile. "I'm sure she is, but are we going to be able to print what she has to say? I'm not worried about us or even the paper as much as I am about Mary. If she is telling the truth like you believe, I'm surprised she hasn't disappeared already instead of just being labeled crazy."

Chad though about that for a moment. "I don't know J.C., maybe Dan's death was enough for them. I think we're the first ones that have asked Mary to tell the story."

J.C. nodded. "Okay, we're gonna run with it. Go on home and type up what you've got. You can bring it in to me in the morning. Bring that tape too. I'll listen to it while you go and have another talk with Mary. How much longer until you finish getting the whole story?"

Shrugging, Chad shook his head. "I don't know at least a couple more days. I think Mary Warren has a lot to share."

*

Chad was in his apartment busy transferring Mary's taped words to his computer. He was just about half-way through his typing when his cell phone rang. Chad sighed, he hated to be interrupted, but his cell phone was his life line. You never knew when the next big story might be coming your way. Chad picked up the phone and frowned when no number or name showed on the screen. Chad shrugged; he supposed there were a lot of people who didn't feel comfortable letting someone else see their name and number, especially if they were calling a reporter.

"Hello, this is Chad Franklin."

Chad could barely hear the voice coming through the phone. "He's not dead."

Chad thought, oh hell, not another crank call. "Who is this and who is the he you're talking about?"

The whispered voice came back on the line. Chad was almost certain it was a male voice.
"Dan Warren, he's not dead. They have him, you have to find him."

Before Chad could answer, the line went dead. Chad sat staring at the phone for a minute.

Why would someone call and say that? Especially now, when he was working on Dan's story? It had to be some kind of hoax.

Somehow word had got out about what he was doing and the crazies were coming out. Although the crazies he knew usually were more than happy to supply a name. In fact they liked seeing their name in the paper. Chad shook his head and sighed. He needed to get his typing done; He pushed the call to the back of his mind.

*

Mary had just finished breakfast and was standing by the one barred window in her room when Ned Findley stepped in through the doorway.

"I need to clear your stuff away before that hot shot reporter gets here." Ned stepped over to the table and began grabbing Mary's breakfast dishes.
He turned and glared at Mary. "No matter what lies you tell him, you ain't getting out of here."

Mary glared back. "Don't worry about it Ned. I'm not trying to get out and I don't have anything to lie about either." Mary turned away and looked outside. The wind was blowing the new green leaves around on the trees. Mary loved the spring; she tried to focus on the nice day.

The only time she was allowed outside was when a nurse was with her. She preferred to just watch the weather from her room. Right now she would rather be doing that than starting off her day being angry just because of Ned's bullshit.

Ned made as much noise as he could taking Mary's breakfast things. When she still didn't look at him, he stomped out of the room.

Mary let out a sigh of relief as soon as she Ned was gone. She sat down at the table and drew deep, cleansing breaths until she felt calm.

Mary was really looking forward to Chad's visit today. As if her thoughts had brought him, Chad came walking through the open doorway.

"Good morning Mary, I hope you're not all talked out from yesterday."

Mary shook her head. "I'm looking forward to it. Oh, and good morning to you."

Chad set his pack full of stuff on the table and sat down. He was tired, he hadn't slept well. After typing up what he had recorded of Mary he couldn't stop thinking about the strange phone call he had received. Then he'd dreamed Dan Warren had come to him or more like Dan Warren's voice had visited his dreams. Which was strange because he wouldn't know Dan's voice if he heard it.

Chad was absolutely sure it had been Dan though. The voice kept repeating, "I'm here, I'm alive, I'm here, I'm alive." Chad had woken in a cold sweat and then tossed and turned as he kept hearing the voice in his head.

Chad turned to Mary. "Come and sit down Mary."

Mary smiled as she walked over and took a seat, then she frowned. "Aren't you going to use your recorder today?"

Chad patted the pack on the table. "I've got it right here. I'd like to ask you a couple of questions first. You know, off the record."

Frowning for a second, Mary finally shrugged. "Okay, what do you need to know?"

Chad's green eyes searched the blue eyes in front of him. The last thing he wanted was to be the one to put a hurt look in those eyes, but until he got a few answers, he knew he wouldn't be able to sleep nights. Chad sighed. "I want to know about your husband's death."

Mary turned away a moment and then back to Chad her eyes glassy. "I told you, it was a car crash."

Chad nodded. "But you still think it was the government that killed him."

Mary nodded. "I know they did. They found a burnt shell that was once Dan's truck. Only the road they found it on was miles away from the route Dan always took."

Chad took a breath, not wanting to ask the next question, but knowing he had to.
"What about Dan? I mean his body?"

Mary shook her head and Chad saw the hurt, haunted look he had wanted to avoid. It was a moment before Mary could speak.

"Gone, burnt, I never got to see him."
Mary hung her head and started crying. It took her a few minutes to finally get herself under control. When she looked up at Chad her eyes were steely blue. "Someone wanted Dan dead. They made sure nothing was left."

Chad shook his head. "I'm sorry Mary, sorry to even ask, but I have my reasons." Chad reached out and touched Mary's hand. "Just one more question."

Mary nodded. "It's okay, go ahead."

Chad sighed. "How did they identify him and who did the identification?"

Mary shrugged. "They sent some remains to a lab somewhere. I don't know where. It was the lab that came back with the determination that it was Dan. To tell you the truth, it was all kind of a blur. Of course it wasn't long after that, I was sent here. First it was emotional trauma, then it was decided I was a danger to myself and others."

Chad stared at her and then he half smiled.
"Did you really pull a gun on Senator Buchanan?"

Chad was surprised when Mary started laughing.

"He came to the house. I didn't know why at first. Said he was representing the state. Telling me how sorry he was about Dan. That he had heard how hard Dan had worked on his invention, how upset he was that Dan's device hadn't been able to get developed. What a boost it would have been for our country." Mary looked at Chad, shaking her head. "Then he started the questions. Was anyone else working on the project? Whose name was the patent under? A lot of questions like that. I didn't like what he was asking. So yeah, I went and got one of Dan's guns and asked the bastard to get out of my house. After he left I went in to the bathroom and threw up. Two days later an official from this hospital and a sheriff came and got me."

Mary just shook her head. "If I wouldn't have done that, I'm sure I would have still ended up here. They didn't want anyone around who might try and get Dan's invention going."

Chad frowned. "What about Trent?"

Mary shook her head. "I don't think he'd even attempt it after all that's happened. Besides remember, Dan left something out of his patent, some small thing, but without it the stack won't work."
Mary cocked her head to one side. "Why the questions about Dan's so called accident?"

Unsure of how, or even if he should share the strange phone call Chad was silent a moment. Finally he nodded to himself. "I got a call last night. I'm pretty sure the caller was a man. The voice was more like a whisper and hard to hear. He said, 'he's not dead, Dan Warren's not dead. You have to find him.' Then the caller hung up. No number and no name on the caller I.D."

Mary took in a shocked breath. Her hand covered her heart where she could feel the accelerated beats. Then it felt like someone had shut off her air. Chad stood and went over to her.

"Bend forward Mary, put your head down. Now slow breaths, easy."

Mary did as she was told, taking comfort in Chad's soothing voice. Finally she sat up. "I'm okay, sorry I just... I..."

Chad nodded. "It's okay, I understand. Sorry, I should have broken that to you better. It was probably just a crank call. It's just, last night I kept playing it over in my mind and I don't know, it felt right, it felt true. What do you think?"

Mary took another deep breath. "I think I wouldn't put anything past the government, including murder. I don't trust them."

Mary laughed bitterly. "Hell, look at where I am."

Chad smiled. "Well, I for one know you don't belong here. You are definitely not crazy."

Mary smiled back at Chad and he could once again see the young woman who a lot of men must have found irresistibly attractive.

"Thanks Chad, you don't know what that means to me." Mary felt like she was going to cry, she was so grateful that Chad had come here and believed in her.

Chad could see how emotional she was. "Listen, I'm going to do some checking. Don't get your hopes up though. Like I said, this was probably just some crazy trying to get some attention."

Mary nodded even though she could already feel the spark of hope growing.

Chad stood up and walked back to his chair and sat down. Opening his pack he drew out his recorder. "Maybe we should get back to the story. I believe Dan and Trent had just met."

Mary smiled gratefully and nodded.

Dan and Mary were driving in his truck, Dan behind the wheel. Mary looked over at him. "I have to tell you, I'm a little nervous about this."

Frowning, Dan turned to his wife. "What are you nervous about? Trent is a great guy."

Mary nodded. "He also so smart his I.Q. is probably way off the charts."

Dan laughed. "Don't worry; he's not pompous about it. You'll like him, I promise."

Mary nodded. "If you say so."

Dan was pulling the truck in to the college parking lot. "C'mon, let's go, Trent will be waiting."

Mary nodded and the two got out of the truck. Mary put her arm through Dan's and they walked together in to the science building. Once inside Dan pointed toward a hallway. "Trent's office is just over there."

A few minutes later they walked in to a small office with the door opened.

Trent turned when he heard them walk up. "Dan, glad you're here. I think I have some good news for you."

Dan smiled. "I sure could use some. By the way, this is my wife, Mary."

Trent smiled at her. "Hi Mary, it's good to meet you. From what Dan's told me, you are one special lady."

Mary shook her head. "I don't know about that, Dan's kind of partial when it comes to me, but it's nice to meet you. Dan has only good things to say about you too."

Trent laughed. "Guess we have something in common then, although I think I may have him fooled. There's a lot of people that wouldn't agree with Dan's thinking about me. However, I do my best to try to avoid those people."

Trent pointed at a couple of chairs. "Why don't you two sit down and I'll tell you what we've come up with."

Dan and Mary took seats and looked expectantly at Trent. He looked first at Dan, then at Mary. "Okay, first let me tell you that in a week we can be set up to do a new test on your invention."

Trent leaned forward. "I promise this time you will get an honest test."

Trent grabbed a piece of paper from off of his desk. "By the way, I checked that water sample you gave me and there is a definite electrical charge. It also looks like it is holding steady."

Trent looked at the paper again. "That was over a week ago and we are still reading that electrical charge, that is very good news."

Trent looked at Dan who was staring back at him in awe. Trent laughed. "I told you I thought I'd find one."

Dan slowly nodded. "I know, but I guess I didn't believe it would really happen."

Mary looked at both men. "You mean that the invention not only takes down pollution but is making energy when it does?"

Nodding, Trent smiled. "Pretty awesome invention your husband has got there. Now we just have to get the tests done and show what percentage of the particulates and gases Dan is actually capturing."

Dan sighed. "This is all a little bit overwhelming and I have to admit I am in way over my head. All I set out to do was invent a way to get rid of the nation's garbage."

Trent laughed. "Well my friend, I'd say you accomplished what you set out to do. Now, if you can have your invention set up in a week, the college can come out and do the testing."

Dan smiled. "Oh, I'll be ready."

Trent stood up. "Great, I'll get everything going here."

Dan and Mary also stood. Dan shook Trent's hand. "I don't know how I'll ever repay you for all you've done."

Trent shook his head. "There's nothing to repay. I want to see you get this thing up and running, God knows we need it." Trent sighed. "I'd love to stay and talk with both of you, but I have a class to teach. The kids get a little crazy when I'm not there. Can we plan on meeting at your place a week from today?"

Dan nodded. "That would be great; do you need me to rent any equipment or anything?'

Trent shook his head. "I think we've got it all covered. If anything does come up, I'll call you, but really I think everything is a go."

*

Dan and Mary left the college in a daze, both trying to wrap their heads around what Trent had just shared with them. Mary looked at Dan as they walked to the truck.

"Let's go sit down somewhere and get a cup of coffee. I'd like to know more about this energy thing. I know you tried to explain what might be happening when I was getting shocked at the house, but it sounds like Trent might have a better idea of what's going on in that invention of yours."

Dan laughed. "He really knows more than me, but let's head over to the diner and I'll try to explain."

Fifteen minutes later the two sat at a table facing each other. The waitress had brought them coffee and half smiled and then shrugged when they had declined menus.

Mary sipped her coffee and stared over the top of the mug at her husband. Dan could see the curiosity in the blue eyes he loved so much. Mary sat her cup down. "Do you think you can explain this to me? What is really going on in that stack of yours?"

Dan raised his eyebrows and then shrugged his wide shoulders. "To tell you the truth, I feel lucky if I can figure out half of what's going on. You know when I first started all of this I was just thinking of using the steam." Dan tipped his head and shrugged.

"I never even told you this before and I probably should have. It just was so strange."

Mary frowned. "What are you talking about?"

Dan sighed. "Some parts of the invention I added because I dreamed about them. Or at least I think I did. I woke up in the middle of the night. I can't even recall the dream I was having, but I woke up thinking, oh yeah, that's how it works. I just knew I needed to add this thing or that one to make the stack work. When I built the invention everything just went together and when I was done suddenly I'm making energy and pulling down pollution." Dan shook his head at his own words.

Mary just nodded. "Actually, that makes perfect sense to me. I've had times when I couldn't figure something out and remember my dad always saying if I had a problem, to sleep on it. When I did that I usually woke up the next morning with the exact solution I needed."

Dan laughed and raised his eyebrows. "If you say so. Anyway, Trent thinks I am somehow changing the oxygen in the water. He thinks I am creating some kind of imbalance in the water molecules. He says when I do the electrons keep spinning and create their own electrical charge. When that is contained in the heavy plastic containers the charge is retained."

Mary just stared at Dan. "This whole thing is amazing and a miracle, to be able burn garbage with no pollution and then to be making electricity."
Mary shook her head. "Just think what that means to our environment."

Dan laughed. "I haven't thought of much else."

Mary smiled and took Dan's hand. "I'm so amazed at you honey. Just remember if all of this gets too overwhelming, I am right here for you."

Dan nodded. "I know and I am so grateful for that. Right now I just can't wait to get this test done and see what Trent comes up with."
Dan smiled. "What did you think of him anyway?"

Mary laughed. "I think I was worried for nothing. I think Trent was a pretty down to earth guy. I also think he's impressed by you."

Dan shook his head. "Trent is light years ahead of me when it comes to intelligence. I have to give him credit though he never makes me feel that way when I am around him. In fact, it's actually just the opposite. He acts like he's learning from me half the time."

Mary smiled. "We're all learning from you Dan. You are light years ahead of everyone in every area."

Dan laughed. "And now who is being partial in their judgment?"

Mary smiled because it was nice to see how humble Dan was. He didn't even realize how very special he was and that in itself was a special quality.

*

A week later Dan and Mary stood outside with Trent. All three stared up at the top of Dan's invention where once again nothing from the fire below was being emitted. Trent was smiling.
"I just know we are going to get a great test out of this."

Dan shrugged, "I hope so, after the deal with Jim Mackey, I have to admit I'm pretty nervous. On top of that, Even if we end up getting a great test, I don't know where to take it from here. I mean I have already been turned down by the State Economic Development office."

Trent was still smiling as he looked at both Dan and Mary. "I might be able to help with that. I spend about a week a month in Washington D.C. I have been asked on several occasions to sit in on conferences. Maybe I can talk to some people and companies, like the E.P.A. for one. After all protecting the environment is their job. I think we should be making them do it. If I have your permission, I'd like to tell a few key people about this invention of yours Dan. Hopefully show them the results of this test and let them know what your stack can do."

Dan stared at Trent. He couldn't believe the offer. "That would really be great. I have to admit I'm totally in over my head here and I would appreciate any help you can give me. When Joel gets the incorporation done, I can

give you a percent of stock in the company."
Dan laughed. "Of course right now, no matter what
percent I gave you it still wouldn't be worth anything."

Trent laughed also. "I'll take it; I have a lot of faith in
the invention and in you."

Three days later, Mary and Dan were home watching the
evening news when the phone rang. Mary and Dan both
stared at the phone sitting on the side table.
Mary finally picked it up. "Hello."

The voice that came over the phone was excited. "Mary,
it's Trent, Is Dan home? I think I have some great news
to share with him."

Mary smiled. "He's right here, let me put him on."

Mary handed the phone to Dan. "It's for you. It's Trent."

Dan nodded and took the phone. "Hi Trent, what's up?"

Trent took a breath. "I hope you're sitting down."

Feeling like his heart had just skipped a beat Dan finally
nodded. "I am, I take it you got back the test results."

Trent was looking at a sheet of paper. "Here's what we
ended up with. The invention is capturing ninety six

percent of the sulfur and ninety two percent of the carbon monoxide. There's more, but the percent's are about the same."

Trent was smiling and Dan could hear the excitement in his voice. "Those results are amazing Dan, they're way higher than what the E.P.A. standards are now. In a perfect world your invention would be put on every incineration process and the standard would be raised."

Dan forced a laugh. "We don't live in a perfect world Trent. What about the particulates?"

Trent scanned the sheets even though he had already been over them a hundred times.
"We got ninety eight percent on the particulate removal. On top of that we still have the electrical charge. That means energy our country is always saying we need, new renewable energy. You are making it with no pollution. I can't tell you how important that is, but I promise next week when I go to Washington I'll be telling a lot of people."

Dan had to smile at the excitement in Trent's voice. He could feel his own spirits rise. "I can't wait until you get back from there."

Trent laughed. "Believe me, you'll be the first one I call."

Dan nodded. "I'll be waiting, thanks again Trent. I couldn't have done this without your help."

Trent shook his head. "I'm sure you would have figured it out."

Dan hung up the phone and turned to Mary who was waiting expectantly. "We're taking down over ninety percent of the gases and ninety eight percent of particulates. According to Trent, that's way over what the E.P.A. has set as a standard."

Mary smiled and then went over and hugged her husband. "I always knew you were a fixer. Now what happens?"

Dan shrugged. "I'm afraid now we wait. Trent will be going to Washington next week and plans on talking to some of the committees back there. Hopefully the government will want to back the building of the stacks and use them on the incinerators that are putting out so much pollution now. We'll just have to hang in there until Trent gets back to see what he finds out."

Mary shook her head. "It's going to be a long week or so."

Dan sighed, "Tell me about it."

*

Mary was home alone when Trent called just a week later. She was in the kitchen mixing up a cake when the phone rang. Mary wiped off her hands then walked over and picked up the phone. "Hello."

The voice she heard was Trent's but the sound of it was totally different than the last time he had called. Mary could tell something was wrong right away.
"Mary, I need to talk to Dan, is he home?"

Mary frowned at the sound of Trent's distraught voice. "Sorry Trent, he's not home from work yet. Can I take a message?"

Trent was in his living room and pacing the floor. "No, I better talk with Dan. When do you expect him?"

Mary turned to look at the clock. "I'd say about an hour." Mary frowned. "Is something wrong Trent?"

Trent sighed. "Yeah, I think something is really wrong. Listen you better not say anything to Dan, I'll call back in an hour and explain."

Mary didn't know what to say, but she knew Trent sounded extremely upset. "Okay, I won't say anything."

Trent stopped pacing. "Thanks Mary, I promise I'll explain everything. I just think I should tell Dan first."

Mary hung up the phone wondering what could have happened. She sat down at the table forgetting about the cake and waiting for Dan.

When Dan came in he took one look at Mary and knew something was wrong. "What's going on Mary? Has something happened?"

Mary looked up startled. She'd been so lost in her thoughts she hadn't even heard Dan come in. She shook her head as Dan sat at the table.

"Mary, I know you better than that, I think you need to tell me what's going on."

Frowning, Mary shook her head. "I don't know Dan, I really don't. I wasn't supposed to say anything but Trent called and sounded really upset. He wouldn't tell me anything, but he should be calling you any minute."

Dan nodded. "I guess we'll just have to wait for his call then. It must have something to do with his trip to Washington."

Mary nodded. "I don't know what else it could be."

Mary looked over at the counter where the bowl sat with the unfinished cake she had started mixing earlier. She shook her head.
"Oh hell, I'm sorry Dan; I didn't even fix anything for dinner."

Dan laughed. "Don't worry about it; we'll dig something up after I talk to Trent."

Mary nodded, hoping they'd feel like eating after the call. She stood up. "I'll just go take a look and see what I have in the fridge that will be fast and easy."

Dan watched his wife. Trent had really upset her and apparently hadn't even shared any information. Dan stood to go and help her when the phone rang. Both Dan and Mary turned to look at it. Dan finally stepped over and picked it up after the third ring. "Hello."

Dan could hear as Trent let out a sigh.
"Thank God you're home. I really need to tell you what happened in Washington. First though I need to tell you I'm sorry. I couldn't get anyone to listen to me Dan. I tried to get them on board with the invention, but it was the strangest thing, usually those people are begging for my input. This time it was a totally different story. I have to tell you something Dan and this is for your own good. Don't try to push that invention of yours right now."

Dan was frowning, not only at Dan's words but also in the rushed way he was talking. "Hold up there Trent. I don't understand this. You were the one so gung ho to get my invention going; now you're saying I should leave it alone? What the hell happened?"

Taking a breath to calm himself, Trent sighed. "Listen Dan, I can't really go in to any details. Believe me that is for your own good and Mary's too. You just have to take my word on this. For your safety and Mary's leave it alone for now."

Dan looked over at Mary who could only hear part of the conversation. She had taken a seat at the table and was staring at Dan, who was shaking his head. "What does that mean Trent?"

The phone line was quiet a moment, then Trent's voice came back on; he enunciated each word he spoke. "It means the government will not let your invention be released." Trent sighed again. "Let it go Dan, for now let it go."

Dan couldn't believe what he was hearing. "This is crazy. What can they do?"

Trent rolled his eyes. "Dan, they can do whatever they want. I didn't want to say anything, but even I was threatened. I am leaving town for a while Dan. You won't be able to reach me. It is better this way. Just put that invention of yours on the back-burner for a while. Promise me that you will leave it alone Dan."

Dan shook his head. "I can't promise you that. Joel has already got a lawyer working on the patent. I'm not going to stop this thing just because a few people in Washington don't like the idea of the invention."

Trent sighed again. "It's more than a few people Dan and these aren't the kind of people you mess with. Believe me, they are capable of anything. Go ahead with your patent I don't think they'll fight you on that. Just don't try and build any more of those stacks. In fact if I was you, I'd probably dismantle the proto-type you have. Please, just give it some time to cool down. I'd hate for anything to happen to you or Mary, especially since I was the one who pushed you and told you to get this thing moving. The one that told you I could do it for you. I'm sorry Dan; I can't tell you just how sorry I am."

Dan was quiet thinking. What the hell had the government done to scare Trent like this? Dan looked again at Mary. If Trent was right, then Dan knew he had to stop working on the invention. There was no way he would ever take the chance that someone would try to hurt Mary.
"Okay Trent, I'll back off. I'm sorry if I caused you problems. How long do you think you are going to be leaving town for?"

Trent laughed. "Maybe forever. Take care Dan and watch out for Mary."

The line went dead. Dan slowly hung up the phone and then stared at Mary who was frowning at him. "What's going on Dan? What's wrong with Trent?"

Dan shook his head as he walked over and joined Mary at the table. "I'm not sure I know. From what I could tell, some of those government people didn't like what Trent had to tell them about my invention. In fact, Trent said he was threatened and I should back off the stack for my own safety and yours."

Mary stared wide eyed at Dan and then shook her head. "We should have known."

Staring at her, Dan frowned. "What do you mean?"

Mary pointed a finger at Dan. "You would be the last one to believe it. You always think the best of everyone, but I don't. I tried to tell you before, our government with their oil and gas industry backings don't want a new energy source."

Dan frowned. "But what about eliminating pollution? The stack could be used for just that."

Mary shook her head. "Maybe they don't want to end the pollution either."

Dan stared at her dumbfounded. "What are you talking about Mary? All the politicians are always screaming about cleaning up the pollution."

Mary shook her head. "It's just talk, think about it Dan. The companies that pollute have to pay a hefty fine to the E.P.A. They are part of the government. You think they want to give up that money they are receiving?"

Dan thought about what Mary was saying. "I hope you're wrong Mary. I'd hate to think our own government would stoop that low, to let pollution continue and put people's health and lives in danger just to line their own pockets."

Dan shook his head. "I need to go and talk to Joel about all this. He already has a lawyer working on the patent. Trent seemed to think it would be okay to finish the patent as long as I didn't try to push the building of the invention for now." Dan shrugged. "Who knows, maybe in a year or two the government's attitude will change."

Mary only shook her head. She didn't really think that was going to happen.

*

The next day, Dan sat in Joel's office. He'd taken the day off work just for this meeting.

Now as Joel sat across from Dan he was shaking his head. "I'm sorry about all of this Dan. Trent called me after he talked to you. I don't know if he told you, but he got fired from the college."

Dan's eyes widened. "No, he didn't say a word about it."

Joel nodded. "Trent didn't go in to any details, but they scared him bad."

Dan shook his head. "He said he was threatened, but he didn't say how or by whom exactly."

Dan stared at Joel. "Are Mary and I in danger, or you?"

Joel laughed. "Not me and I'm sure you and Mary are okay. You are going to have to put this whole thing on the shelf for a while though."

Dan frowned. "What about the patent?"

Joel shrugged. "We'll finish it. As long as it remains just an idea on paper and not an actual work in progress you'll be fine. When the patent's done, you'll have yourself some official documentation and also some protection against someone trying to steal the idea. Then someday when the time is right, you can try again to get it developed."

Dan nodded. "Thanks Joel, I feel a lot better. I just hope out government comes around to the right way of thinking. I hope they can put people ahead of money and power."

Joel nodded. "Me too Dan, me too."

Chapter 9

Chad reached out and turned off the recorder. "Sorry to stop you Mary, but I just have a couple of questions before you go on."

Mary smiled. "Just a couple?"

Chad laughed. "Well, let's start with a couple."
Mary nodded and Chad continued. "What happened to Trent?"

Mary shrugged. "We never talked to him again. He disappeared for a couple of years and no one ever knew where he went. Then we heard he had shown up at another college. But then we found out he had gotten fired from there for writing a controversial pamphlet and passing it around the campus."

Chad frowned. "Controversial how?"

Mary sighed. "He blamed the President for the whole nine eleven business. He thought the President was behind the entire thing. I didn't hear any more after that. For all I know he could be dead too. Whatever happened in Washington when he tried to help Dan ruined his life and scared him bad."

Chad nodded. "What about that Senator? He's the one who came to your house wasn't he?"

Mary nodded. "At the time I didn't realize he was the one pulling the strings. Trent had told Joel that a Senator was the one stopping Dan's invention. It was only after I ended up here that it dawned on me that Senator Buchanan had to be the one. That's why he came to me and asked me all the questions and I am damn sure that's why I am here."

Chad frowned. "But if Dan put the invention away, why did they kill him?"

Mary sighed. "He only put it away for a few years. He kept trying to push it over and over again. He just wanted to help." Mary shook her head with tears in her eyes. "Dan just saw every day on the news there was more and more talk about the damage pollution was causing. He felt he had to try, no matter what."

Chad was going to say something when his cell phone rang. He looked at the number and then at Mary. "It's my boss, I better take it."

Chad walked out in to the hallway and answered his phone. "What's up J.C.?"

The voice on the line sounded upset. "I need you back in the office Chad."

Chad frowned. "I'm still doing the interview with Mary."

J.C.'s voice was louder than normal. "No Chad, you're not. Get back to the office now."

Chad sighed wondering what was going on. "Okay J.C., I'm on my way, just let me give Mary an excuse and then I'll be there."

Chad stepped back in to the room. "I'm sorry Mary; we are going to have to cut the session short. Something's come up and J.C. needs me back at the office."

Mary nodded and then looked at Chad with worry in her blue eyes. "You'll be back though won't you?"

Seeing the look on her face, Chad felt like a piece of his heart had broken away. He nodded. "You couldn't keep me away."

A slight smile appeared on Mary's face, but Chad could still see the worry beneath it and hated that he had been the one to cause it. He hurried and gathered up his things. "I'll try to get back later today, but if I can't make it, I'll be here first thing in the morning, promise."

Mary nodded and watched Chad walk out the door.

Ten minutes later Chad stepped into J.C.'s office. "What's going on J.C., I was right in the middle of Mary's interview."

J.C. looked up at the young man. "Sit down Chad. We need to have a talk."

Chad pulled out a chair.

From the look on J.C.'s face Chad was sure he wasn't going to like what his boss had to say.

J.C. rubbed his temples with his fingertips, like he was trying to rub away a headache, or maybe just the thoughts he had to share. Looking at him, Chad figured it might be a little of both. Finally J.C. put his hands down on his desk. "I'm going to have to ask you to stop working on this story of Dan and Mary Warren's."

Chad just stared. "What are you talking about J.C.? You were the one all gung ho to have this story finally come out."

J.C. held up his hands. "Calm down Chad. I'm doing this for your own good and more important for Mary's."

Chad glared at J.C. "Who got to you?"

J.C. shook his head. "This story is trouble and I mean big trouble. If you even tried to print it, then it would be the last thing you ever printed."

J.C. leaned on his desk. "Chad I'm serious about this. There are some big names out there who don't want this story to come out."

Chad shook his head. "Jesus J.C., I think you're scared."

J.C. nodded. "You bet your ass I am and you should be too."

Chad sighed. "What about Mary? She finally has someone willing to listen, you want to stop that? It's more than that J.C., I believe her, I believe every word damn it.

J.C. nodded. "I know you do Chad, so do I. There's nothing either of us can do right now."

Shaking his head, Chad frowned. "I need to go talk to her."

J.C. shook his head. "I wouldn't advise that Chad. Your best bet is just to forget about Mary and the story."

All Chad could do was stare at his boss in wonder. He had never seen J.C. scared before, didn't think his boss was capable of that emotion. He didn't like that he was seeing it now. Chad pushed back his chair and stood up. "I don't know what's going on, but I think I have a pretty good guess who is pulling the strings. I think I need to go home and sort a few things out."

J.C. nodded. "Maybe you should. Just remember what I said. I know you like Mary and that's why you have to think what's best for her."

Chad nodded, that's all he was thinking of. He turned and left the office. He knew he had to go back and see Mary, but first he was going to go home and do a little digging. Chad left the building still trying to figure out J.C.'s attitude.

He'd seen J.C. take on a lot of controversial topics and to see him now so unnerved wasn't helping Chad's own nerves any. Then there was that strange phone call. Add all of that to Mary's unique story and things didn't look so hot.

Chad made it to his place and went straight to his computer. The thought crossed his mind that maybe people might be watching him or even monitoring his computer, either way he still had to do this. He couldn't stop now. He knew he was getting close to discovering something big.

Chad looked up Dan Warren's patent first. He was relieved to see it was still listed at the patent office and it looked like it didn't expire for a while. Someone had to be paying the fees. Chad thought maybe it was Joel and made a mental note to ask Mary about it when he saw her.

Next he looked up Trent Altman. There was quite a bit on him, but nothing after the nine eleven thing Mary had told him about. Chad didn't like the looks of that. Where was Trent? Chad sat back and stared at the computer screen. He wanted to try one more thing, then get his butt over and talk to Mary. He'd decided he wasn't going to tell her that J.C. had taken him off the story. He still was in shock about that himself. He hadn't expected J.C. to kill this story.

Chad sighed as he typed in Senator Buchanan's name for a Google search.

After a half an hour the only dirt on Senator's Buchanan's hands came from a few unsavory tabloids. Chad shook his head, there had to be more than that. He knew the Senator had at least one failed bid for the Presidency. Usually when someone ran and got knocked out of the race it was because someone found some skeletons in your closet. According to Mary, Senator Buchanan was at least one of the participants in suppressing Dan's invention. Chad believed her. He didn't know much about the Senator, but politicians in general were good at cover-ups and they didn't care who they hurt in the process.

Chad wearily shook his head and then shut down the computer. He still knew a few people who might be able to shed some light on all of this.

Chad left his place and drove to Rose Hills. When he got there he went straight to Mary's room.

Mary looked up from the table where she was sitting when Chad stepped in and her face lit up. "Chad, I wasn't sure you'd be back."

Chad joined her at the table. "I told you I'd be back; I always keep my word, especially to a beautiful woman."

Chad stared at the face in front of him. "Listen Mary, I am not going to let you down. My boss just asked me to help cover a big story we're doing. We'll just have to switch our time around a little."

Mary nodded and then frowned. "Where's your recorder?"

Chad shrugged. "I left it back at the newspaper office. I thought we could just talk. I'll be sure and bring it with me tomorrow. It will probably be in the evening though. That is if it's okay with you."

Smiling with relief, Mary nodded. "That'll be great, thanks Chad."

Chad shook his head. "Don't thank me, this is one story I think needs to be told." Chad frowned as he thought about J.C. and what he had told him. He didn't want to think about that now; instead he looked puzzled at Mary.

"By the way, I looked up Dan's patent and it looks like the fees are paid up for another eight years. Are you the one paying on that?"

Mary shook her head. "No, to tell you the truth with all that's happened I never really thought about the patent."

Chad shrugged. "Maybe Joel is taking care of it then."

Biting her lip, Mary shook her head. "Joel died two years ago."

Feeling like his stomach dropped and almost afraid to ask, Chad stared at her. "From what?"

Mary looked down at the table and back at Chad. "He had cancer and died a slow, agonizing death. It was really bad."

Chad shook his head. "I'm sorry; I know you and Dan thought a lot of him."

As she tried to smile, Mary nodded. "We did, Joel helped us so much. Not just on the invention either, he was a good guy." Mary thought…and good guys finish last, just like Dan.

Chad shook his head. "That still leaves us wondering who's paying the patent fees."

Mary looked at Chad. "Is it important?"

Chad shrugged. "I guess not. Dan may have paid them ahead of time himself."

Chad leaned forward in his chair. "What else can you tell me about Senator Buchanan?"

Mary sighed. "Not a hell of a lot. I just know Trent said he was the one pulling the strings. Senator Buchanan talked to both Jim Mackey about the testing and Ron Letner at the State Development Center. I'm sure you couldn't get either of them to admit to that though. He has to be holding some pretty big threats over everyone's heads."

Chad was silent a moment thinking. "Did you say Dan tried to get the invention going a few different times?"

Mary grunted and then nodded. "Yeah, and always with the same results. No matter where he went something always happened to stop the invention. He even had investors that came up hurt or missing. I almost began to feel like the whole thing was cursed."

Chad stared at Mary waiting. He knew she had more to add. Mary closed her eyes for a minute. When she opened them, Chad read the hurt and sorrow in their depths. She shook her head slightly.
"Some people say it's all just a coincidence, but that just doesn't cut it in my book. Too much has happened along this crazy journey for me to buy that. I won't go in to everything, but I'll share a couple of the stories."

Chad nodded and studied Mary's face as he waited for her to continue. Mary nodded, more to herself than to Chad. "The first thing that happened I remember I really did think was just a coincidence. A group of investors with a lot of money were all fired up about getting Dan's invention going. They had met with Dan and Joel and set up some kind of agreement. I don't have the exact figures, but it was over fifty million. Two weeks after their meeting, the whole group was involved in a financial crash. According to what Joel told Dan, these investors lost everything. Dan was so upset when he

heard about it he was physically sick. The next thing
was a physics professor. He was going to lend his name
to the project; you know just to give it more credibility.
Anyway, he got in a car wreck and broke his neck. He
didn't die, but he is a quadriplegic. Another investor had
his wife die from unknown causes; He was so devastated
he walked away from the company. There's quite a long
list of the so-called coincidences, but I think you get my
point."

Shaking his head wearily, Chad nodded. "Yeah, one or
two might just have been bad breaks, but after a while
you have to stop and say what the hell is going on."

Mary nodded. "That's why Dan finally just stopped
trying. That was a year or two before he passed away.
That's why I can't understand why this cover-up is still
going on."

Chad shrugged. "I don't know, but like I said the patent
is still good."

Mary nodded. "Maybe it is, but no one can build it and
make it work. Remember Dan kept something hidden.
He left that key ingredient out of his patent. Without it
the stack won't function the way it is supposed to."

Chad frowned. "The Government must know that.
Maybe they even tried to build one of Dan's stacks and
it didn't work. They had to have known how well his
invention worked from the testing and from that they
would have figured out that the patent didn't have
everything in it."

Chad's mind was whirling. Could the Government really have faked Dan's death? Was he alive and being held somewhere? Chad rubbed his forehead trying to slow down his mind. He looked at Mary and knew she had to be thinking the same things. Chad wanted to tell her it was best down not head down that road, but couldn't because that's exactly where he was headed. Chad reached across the table to take Mary's hand to calm her when the door opened and Ned burst in. He had a sneer on his face.

"I don't think holding hands was part of the agreement when you were allowed in here Chad."

Chad stood up and glared at Ned. "Not that it's any of your business, but some people can be friendly without having their filthy minds make it into something it's not. I think you'd be better off minding to your own affairs Ned."

Ned looked like he was going to explode. His eyes were open wide and his face had turned red. His hands clenched in to fists. "Maybe Mrs. Delaney might not think the two of you are so innocent."

Chad shook his head. "Whatever Ned, I think the administrator of this hospital is too smart to buy into your crap. Now, if you don't mind, Mary and I were talking."

Ned turned and stomped out of the room.

Mary laughed when he had gone. "Don't worry about Ned. I don't think anyone likes him and I doubt if he would even go and try to talk to Mrs. Delaney. She'd see right through his little tantrum anyway."

Chad nodded. "You're right, but I sure don't like that guy."

Mary nodded in agreement then looked up at Chad. "Any more questions?"

Chad shook his head. "No, to tell you the truth I think we are going to have to call it good for tonight. I really need to get home. Mostly I just wanted to come by and let you know about the change in times. Will it be okay if I come by in the evenings from now on? I can probably get here right after you have dinner, if that sounds okay."

Mary stood up. "I'd really like that. So, I guess I'll see you tomorrow night then?"

Chad smiled. "It's a date and I'll remember the recorder."

Mary watched as Chad walked out the door. Once again she felt alone and sad.

*

Chad left the hospital and stopped by the liquor store to grab a fifth of vodka. He didn't drink a lot, but knew he would be needing it for tonight.

One way or another he was going to find the truth behind what happened to Dan and why Mary was in that place. He had a few phone calls to make and knew he would need some extra fortification for them.

When Chad got home, he mixed himself a drink and sat down with it in front of his computer. Chad tried looking at every liberal newspaper he could think of for any kind of dirt on Senator Buchanan. An hour of searching still didn't give him much.

Chad sighed and picked up his cell phone. He knew one person who could maybe get him some answers. Much as he hated to, he dialed Ernie Ashford's number. Ernie was a freelance writer who seemed to know the worst about everyone. Chad hated to admit it, but Ernie's stories were accurate nine times out of ten. Ernie didn't care who he hurt when he wrote the dirt he had dug up. All he wanted was credit for the story and he didn't care what innocent bystanders got hurt. Chad had never liked Ernie although their names had been linked together on at least two articles for the paper. Chad had agreed to share the stories because truth be told Ernie had been the one to crack the stories.

One had been about a drug pusher and the other a chop shop that dealt in hot cars. Both men involved had used young kids to do their dirty work. Chad had wanted to nail them so bad he had gone in with Ernie. The plus side was that both men had ended up with jail time; the minus side was working with Ernie. Chad just couldn't find a way to like Ernie. First off, the guy was a pig. His greasy black hair always looked like it could use a good washing, right along with Ernie's dirty clothes. He favored wrinkled dress pants and usually a dirty t-shirt. Neither article of clothing did much to conceal the fat tire of flab that hung over Ernie's waist band. It was more than that though. Somehow Ernie was able to find the worst skeletons in someone's closet and dig them out. Chad was okay with that when it had been people like the drug pusher and the auto thief, but he'd also seen Ernie hurt people that he had no business bothering. Innocent victims were just as much on Ernie's hit list as the bad guys. That was one thing Chad just didn't have the stomach for. Right now though, he needed to find out what the skeletons were in Senator Buchanan's closet. He needed to help Mary and if he had to use Ernie to do it, then he would.

Ernie answered on the second ring. "Ashford here."

Even the sound of Ernie's voice irritated Chad. He sighed. "Hey Ernie, it's Chad. I was wondering if you could do me a favor."

Chad reached over and took a big swig from his drink, wishing he would have had a few more before he had made the call.

Ernie laughed, but it sounded more like he was grunting. "Well now Chad, I'm sure we could work something out. What kind of favor are we talking about exactly?"

Chad felt sick just listening to Ernie's voice. He took another big drink from his glass. Even though he had mixed the vodka with orange juice, he could still feel the burn of the liquor as it went down. Chad savored the burn, it was better than the queasy feeling he got from Ernie. "It's like this Ernie, J.C. is sure that Senator Buchanan is going to make another run for the Presidency. He thinks Buchanan may even end up being the front runner. Before that happens, we'd like to get some background on him and to tell you the truth, I'm coming up pretty much empty." Chad hoped nothing in his voice would betray the fact he was lying.

Ernie laughed. "I think I can help you there Chad, of course I'd want my name on the article, for starters."

Chad rolled his eyes. "What else besides that?"

The phone went silent a moment.

"Let me think on it. I'll get the stuff on Buchanan ready and give you a call tomorrow night. By then I'm sure I'll have something to add. Don't worry I'm sure I can think of a good trade for my information."

Chad wasn't liking the sound of this, but reminded himself how bad he needed the information.
"I'll talk to you tomorrow then."

Chad pushed the end button on his phone and threw it down on his computer desk. Then he wiped his hand on his pants.

Just talking to Ernie made him feel dirty.
Chad took his now empty glass to the kitchen for a much needed refill. He decided he better grab some food while he was there. He needed to give the vodka something to sit on. Chad made a sandwich and grabbed a bag of chips. He took those and his drink in to the living room. He wanted to catch up on some news. Chad picked up the remote and turned on his flat screen TV. A reporter came on the screen talking about tornadoes in the Midwest. Chad flipped the channel only to see another reporter talking about the severe weather that was hitting the northeast. It seemed every place was having more severe and unusual weather than ever before. Chad wondered if Dan's invention would have changed all that if his idea wouldn't have been suppressed. He sighed and flipped off the TV. Maybe he didn't really want to watch it after all. Chad finished his dinner, if you could call it that and then made another drink. He hoped he wouldn't be paying for it in the morning, but it was just one of those nights when a few drinks filled the bill. Chad put his legs up on the coffee table and tried to relax. A task he was finding impossible. His mind was restless as he kept thinking about J.C.'s warnings about Mary and the story. Who had gotten to him? Chad couldn't imagine what kind of

threats had to have been made to scare a man like J.C. Chad kept seeing Mary's face and knew there was no way he could stay away. Somehow he had to find out the real story.

He hoped the Ernie could come up with something on the Senator. Chad was sure he was behind everything, he had to be. Chad closed his eyes and tried taking several deep breaths. He finally got his thoughts to slow down when his cell phone went off. Chad considered just letting it ring, then shook his head and went over and grabbed his phone up from off the desk. He looked at the I.D., no name and no number. Chad felt his heart speed up.
"Hi, this is Chad Franklin."
Chad's rapid heartbeat almost came to a standstill when he heard the deep whisper.

"You need to hurry."

Chad almost yelled back in to the phone. "Who the hell is this?"

The whispered voice came back. "Check the old mines, he's there."

The line went dead.

Chad shook the phone in frustration. Then he sat down to think. He wished he hadn't had that last drink. He should make coffee, but then he would never get to sleep. Chad put his head in his hands and sat still as he waited for his thoughts to clear. He wanted to focus on the call and the caller's few words. The voice had said to check the old mines. Chad knew south of town there were abandoned mines. When other countries started shipping in steel at a cheaper price, the mines had closed down.

A lot of the workers had also saw their retirement savings go down the drain too.

Chad got up and went over to the computer. The next hour was spent going over the mine's history. From what he could tell, no one was using the mines now. Could Dan really be there? Chad looked at his watch and was surprised to see it was close to midnight. He could drive out there, but thought he had better wait. First off he had been drinking and he also wanted to talk to Ernie. Also he felt it would be better for him if or more like when he did go out to the abandoned mines; he let someone know about it, just in case.

Chad turned off the computer and headed for his bedroom although he didn't think sleep would come easily with all that had happened.

The alarm going off the next morning proved Chad wrong. He couldn't remember anything after he'd lain down and his head had hit his pillow. Maybe the vodka had been good for something after all. Chad reached over and turned off the alarm. Then sat up in bed and stretched. He was pleasantly surprised at how rested he felt and at the fact that he didn't have a headache. Chad was also grateful because he knew he'd need all the energy he could get today.

An hour later, Chad walked in to the newspaper office and headed straight to J.C.'s office. He knocked on the glass doors.

His boss looked up and waved for Chad to come in. "Chad, it's good to see you. I've got a couple of things I'd like you to work on."

Chad nodded. "Okay, but if it's alright, I'd like to be out of here by four o'clock today."

J.C.'s eyes narrowed. "You're not up to something you shouldn't be are you?"

Shaking his head, Chad smiled. "C'mon J.C., you know me better than that."

J.C. nodded. "That's why I'm worried. I know you Chad; you're a sucker for a sad story or a pretty face. Mary Warren just happens to fit both categories." J.C. looked down at the floor and shook his head before looking back at Chad. "I'm sorry that I can't tell you anymore then I have, but be careful Chad. If you try and help Mary, things are gonna go very wrong, very fast."

Chad tipped his head, thinking. "I get it J.C. and don't worry I'll be careful. Really, I just want to stay friends with Mary. It seems to me she's a little short in that area. What really pisses me off is that wasn't any of her doing."

J.C. sighed and walked around his desk to pat Chad on the shoulder. "I understand how you feel. Right now how about I put you to work, that should keep your mind busy anyway."

Chad nodded and got his assignments from J.C. and got busy doing research and typing up some stories.

Chad was surprised when he looked up at the clock and saw it was almost four. At least J.C. had managed to keep him occupied that long. Now he had to worry about the rest of his day and night. Chad went and found J.C. to let him know he was leaving for the day. J.C. looked at him. "Do me a favor and keep your cell phone on. If you get in trouble, give me a call. I don't know how much help I can be, but I'll damn sure give it my best shot. I don't like all that's been going on either Chad."

Chad nodded. "I told you J.C., I understand. There's no need for you to get in trouble or put the paper in jeopardy."

J.C. smiled, "Thanks Chad, that really means a lot."

Chad left the office and headed home. Before he got there his phone rang. He had a hands free attachment and his caller I.D. was set up to read out on his radio screen. "Hey Ernie, what's up."

On the other end of the line Ernie was smiling, but the smile wasn't quite normal. It was more of a cross between a gloat and a sneer. "I think I've dug up a few things you can use."

Chad was leery, but he really needed the information. "Great, can you call me back in ten minutes? I'm almost home and I want to write this down."

Ernie laughed. "Sure thing, what I've got you'll be glad to have."

Chad ended the call and finished driving home. When he got there he went inside and grabbed his notebook. He thought about trying to record the conversation, then changed his mind. He didn't think Ernie would be able to tell, but he didn't want to take any chances either.

A couple minutes after Chad had been home, his phone rang. Chad picked it up after the first ring. "Okay Ernie, I'm ready whenever you are."

By the sound of Ernie's voice, Chad could tell he was eager to talk about what he had found. Probably because he knew Chad had come up empty handed. "The first thing you should know about is the old mines."

Chad frowned as his heart skipped a beat. "You mean the ones south of town?"

Ernie sat back in his chair with a grin of his face. "Yeah, it seems the good Senator sold out the guys out there. He's made a deal with the Chinese. Of course you won't find his signature on any of the paperwork."

Chad nodded, he'd already figured as much. "What are the Chinese going to do out there?"

Ernie grunted. "What do you think? They're going to start up the iron mines, full production from what I hear. They will be bringing in their own workers of course."

Chad felt sick thinking about all the workers that had been let go when those mines had shut down, all the families hurt. Then he thought about the whispered voice. "Are they already out at the mines?"

Ernie's voice grunted. "Listen, you really can't say much about this yet, but my sources tell me there's a skeleton crew out there at night. I've talked to a few people who swear they seen 'em."

Chad's mind was whirling again and he knew he was going to be making a trip to the mine. "Okay Ernie, what else you got?'

Ernie talked for about a half an hour and Chad took notes, but all he could think about was the mines and the chance that Dan Warren could be alive and being held out there.

Ernie's voice broke in to Chad's thoughts. "Hey, you still there Chad?"

Chad shook his head to clear it. "Yeah, sorry, just thinking. Can anyone collaborate any of this?"

Ernie laughed. "What, don't you trust me?"

Chad knew how he'd like to answer that question. "It's not that Ernie. You know I can't do a story without verification."

Ernie's voice sounded upset. "You get ready to print the story and I'll have your verification. Just don't forget I want my name on it and you still owe me. A lot of this is big stuff."

Chad rolled his eyes. The last thing he wanted was to owe Ernie Ashford anything. He had to admit Ernie had come up with a hell of a lot more information on the Senator than he'd been able to. "I won't forget Ernie, I'll be in touch."

Chad hung up the phone realizing he hadn't thanked Ernie for the information he shared. Chad felt like he'd sold his soul to the devil and he knew Ernie would get his thank you in the form of whatever favor he came up with. Chad also knew that whatever it was, he'd deliver if he could. Chad always believed in paying your debts, no matter what form they came in.

Chad stood up and paced the floor. He needed to go and see Mary. Until he actually went out to the mines and saw what was going on for himself, he didn't want to share what he suspected with Mary. She'd already had enough hurt and disappointment. He didn't want to get her hopes up and then dash them again. Just because he'd gotten that crazy phone call and the story from Ernie, there was still no real reason to believe Dan could be alive. What kind of sick person would do that anyway and why? Chad stopped pacing and sighed, one thing at a time. Chad grabbed his pack with the recorder and his notebooks and headed over to see Mary.

He found her staring out the barred window. She turned when she heard him enter, her face lit up with the smile she gave him.

Chad could still see the red eyes and knew she had been crying. His heart ached at the picture he saw before him. He stepped toward her.

"What's wrong Mary, what happened?"

Shaking her head, Mary took a deep breath. She didn't want to start crying again, not now. Chad was here and everything was going to be okay. "I'm okay, really, I was just…"

Mary stopped speaking as her voice cracked.

Chad took her hand. "C'mon and sit down and then you can tell me what's wrong. I know something has happened."

Mary let Chad lead her to a chair at the table. She sat down and tried to smile. "I'm just glad you're here. I feel better now."

Chad nodded. "Just take your time."

Mary sat for a minute twisting her hands in her lap and staring at them. Finally she looked up at Chad. "I spent three hours in with my shrink today. I'm almost certain that was Ned's doing. I didn't think he had the nerve to actually go and talk with Mrs. Delaney, but it had to have been him. The psychiatrist is questioning whether your being here is in my best interest. He is trying to say that us talking about Dan has somehow caused me undue suffering and has set back any recovery I have made." Mary shook her head. "Like they have ever been worried about helping me make a recovery, no way in hell do they want me leaving this place."

Shaking his head, Chad looked out raged.
"Jesus, what kind of idiotic thinking is that? I thought recovery was supposed to be talking about your feelings. Maybe I should go down and have a talk with Mrs. Delaney myself."

Mary shook her head. "No Chad, that would only make things worse. I'm sure I convinced the psychiatrist that talking to you has helped me. I was just upset about the way they handled the whole thing. Then I talked myself in to thinking you wouldn't show up tonight. I'm sorry Chad, I shouldn't have worked myself up like that. It's just that you're the only one I've talked to in so long. Other than that damn psychiatrist and one or two of the nurses and that's not the same. They just think I'm crazy and would like me to stay that way. I don't talk to them about Dan or what happened. Besides all of that I really think you actually believe me."
Mary looked at Chad, wide eyed. "You could never know how much that means to me."

Chad smiled. "The feeling is mutual Mary. What you've been and are still going through."

Chad shook his head. "You are one special lady. I have to say you are about the most interesting and strongest person I know. Don't worry about Ned or that shrink; I won't let them stop me from coming. Now, do you have more of Dan's story to share or do you just want to talk?"

Mary shrugged. "I don't think there's much more to add to Dan's story. He finally got disgusted and just quit trying to get his invention going."

Chad leaned forward. "Mary, do you think you could tell me about the day of Dan's accident?"

Chad could hear the sound of Mary drawing a deep breath.

She nodded slowly. "I think so, but there isn't really much to tell. Dan worked that day. He was late getting home, and I was starting to worry. Dan always tried to call me when he was going to be late. He didn't have a cell phone, I know a lot of people carry them, but Dan never carried one. He always said he'd been born in the wrong generation. I waited almost two hours and then tried to call his work, but no one answered. I finally called his boss at home. He said Dan had left on time for once. That was when I really started worrying. I thought about going to look for him, I knew Dan always drove to work and back home using the same route. I decided to wait just a little longer. When I finally couldn't stand it any longer I grabbed my keys and got ready to go look. About the same time I got to the door the phone rang. I ran to answer it; I was just sure it was Dan."

Mary shook her head. "It wasn't, it was a police officer saying there had been an accident and they were sending a patrolman to the house. He showed up about ten or fifteen minutes after that call with the details on the crash. Dan's burnt truck had been found. They said his license plate had fallen off and that's how they had identified the wreckage, nothing else but ash was left."

Chad frowned. "That seems a little strange and pretty handy for them."

Mary nodded. "I was so upset at the time I didn't give it much thought. I just knew that my Dan was gone. I guess you can add that wreck to the list of other coincidences that happened. It wasn't too long after that wreck that I ended up in this place."

Chad shook his head. "I am so sorry for what you've been through Mary. I would love to be able to just take you out of here. In fact it would probably be the best thing that happened if I could pull it off."

Mary laughed. "I don't think the administrator would let you do that, or Senator Buchanan for that matter."

Chad looked at Mary with a slight smile. "You never know Mary, you just never know."

Mary frowned at Chad's words, but somewhere deep inside her also felt a glimmer of hope.

Chad shook his head. "Let's talk about something else. Why don't you tell me about you and Dan before this invention entered your life?"

Smiling at him, Mary nodded. "I would love to do that Chad."

Mary talked for over an hour recounting her life with Dan. Chad was surprised and pleased to hear Mary laughing as she recalled their good times together. He could tell Mary and Dan Warren had loved each other very much and seemed to have had an almost perfect life until Dan had stepped on the wrong toes with his invention. Chad hated to end the conversation, but he had a lot to do. He pointed at his watch.

"I'm sorry Mary, I would love to stay here and listen to more, but I'm going to have to get going."

Mary nodded. "It's okay, it's better if you don't stay longer. I don't want to end up in another three hour session just because you might have spent a few minutes too long here."

Chad laughed. "I hear you there. Is the same time tomorrow okay for me to come by?"

Nodding her approval at Chad, Mary then frowned. "I'm done telling the story though."

Chad shook his head. "Don't tell anyone that. As long as they think I'm still getting information from you for the article they'll let me visit. Besides, I'm sure there are a lot of items you haven't shared yet."

Mary smiled and nodded. "Okay then, I'll be ready to continue tomorrow evening."

Chad gathered up his things. "I'll see you tomorrow Mary, sweet dreams."

Mary laughed. "From your lips to God's ear Chad, I sure could use them."

Chad walked out, hating to leave Mary behind. He really needed to figure out a way to get her out of that place. No matter what he discovered about Dan, Chad still felt that Mary's days were numbered. Thinking of Dan, Chad knew what he had to do next. He left the hospital and went to his car. Chad sat in his vehicle thinking. He knew he should call J.C. and at least tell him that he planned on going out to the mines. That way if he didn't show up for work in the morning J.C. would know why. Chad pulled out his cell phone and stared at it. After a few minutes he laid it on the seat beside him. If he had problems he'd make the call. Right now he felt it was better not to involve J.C. Chad started his car and headed out toward the mines.

It took Chad about forty five minutes to reach his

destination and he was still about a half mile away from the main mine building. He could have went in on the main road and cut his time almost in half. The main road was old and rutted but looked like a new major highway compared to the road Chad had decided to use. The road he chose was more dirt and large rocks than asphalt, but Chad felt safer using it. He grabbed his cell phone and a small flashlight out of his car. There was still enough light coming from the nearly full moon that he could see where he was going. Chad was more concerned with what he couldn't see hiding behind the numerous trees and rock outcroppings.

Chad slowly made his way to the mine buildings, keeping his eyes and ears open. Stopping each time he heard something scurrying into the bushes. This place was a haven for jack rabbits and deer.
Chad approached the main building from the back side. Even though he believed Ernie's story he still was surprised to see three cars and two pickups parked behind the building. Most of the windows had been boarded up when the mine had shut down, but Chad was still able to see faint light coming through the cracks. He crept to the side of the main building where most of the light was coming from. Chad crouched beneath one of the boarded up windows. Voices floated out to him, but Chad couldn't understand the words. It took him a moment to realize that was because the words weren't in English. At first he had mistakenly thought the miscommunication was because he could barely hear the muffled words being spoken. When he did recognize the

true reason his heart skipped a beat. Although he didn't speak or understand a word of Chinese he was almost certain that was the language that was being spoken inside the mine building right now.

Chad leaned his back against the wall as he hunkered down thinking. He figured if Dan was being held it would be here in the main building. Staying low, Chad made his way slowly around the building looking for some kind of entrance he could sneak in to. On the opposite side of the building he found a basement window that opened inward.

Chad applied slight pressure to the window and sighed with relief when it opened.

Chad pushed it open as far as it would go, then put his head in the opening and listened. He could hear two people talking and this time the voices were definitely in English. Chad leaned in further so he could look around. His body was tense, ready to jump back if he saw or heard anyone. Chad let out the breath he had been holding when he didn't see or hear anyone in the area. From what he could tell inside the dark room it looked to be an open storage area. From the amount of dust and dirt it was also an unused one. Chad could see there was enough room for him to squeeze into the window but he would have to crawl on to the window frame to do it, and then try to drop from there to the floor below. He was afraid if he put his weight on the frame both he and the window would come crashing down. Chad stuck his head in and looked at the wall on either side of the window. On the right side there was a large hook with a few old chains hanging from it. Chad nodded to himself. He was sure he could use that to get down. He just hoped he could find something to stack under the window so he could get back out. Chad slowly and carefully climbed through the opening. Reaching over he grabbed the hook and swung his body in. The chains clanked against the wall with the motion. Chad froze waiting for someone to come running in to the room and grab him. When no one did, he dropped to the floor.

Turning to face the room, he leaned back against the wall holding his breath. He stayed that way for a minute. When he was sure he was still alone, Chad let out his breath. He looked around the room. In one corner, Chad found a large wooden spool that must have been used to wind up large cable lines. The spool was empty now and Chad pushed it over under the window. Chad looked at it and nodded. If he made it back here, he would at least have a chance of getting out.

Chad walked in the direction he thought the voices were coming from. Across the storage room he saw a hallway and walked down it, glancing nervously behind him with each step he took. The hallway led to another large room like the one Chad had just left. At the end of the room a large cage about the size of a jail cell sat. The bars went from ceiling to floor and spanned all the way across the room.

As Chad looked he saw two men standing in the makeshift cell. The taller of the two men was talking.

"Hank, I think you better do just what they ask."

The other man shook his head of gray hair. Both of the men had fairly long beards, although the tall man's was an inch longer than the other man's.

The shorter man answered. "Damn it Dan, I'm not doing it. I am not going to help them steal jobs from people around here. If they want to reopen this plant then they should be hiring back the original workers. I know all of those people Dan and a lot of them lost everything they had. This is bullshit anyway, locking me up in here until I agree to help them. And look at you, what did you ever do? There's no way in hell you should be here. Another good reason I shouldn't cooperate."

Dan shook his head. "Don't worry about me. If you cooperate at least you could get out."

Hank laughed sarcastically. "Are you kidding? After what I've been through and seen out here? There's no way in hell they are going to let me out no matter what I do."

Chad stepped into the room, putting a finger to his lips as he did. Both men looked over as they heard Chad's footsteps expecting to see one of their captors. Both men frowned as they saw someone totally different standing in front of them than what they had expected. Hank started to say something, but Chad waved a hand stopping him. Chad turned to look at the other man.

His voice was a whisper. "Are you Dan Warren?"

Dan nodded. "But who the hell are you?"

Chad answered in a more normal tone. "My name is Chad Franklin. I know your wife. I'm a reporter and have been interviewing her."

Dan frowned. "Mary, why would you interview Mary?"

Chad shook his head. "It's too long of a story to go in to now."

Dan's face softened. "How is Mary?"

Chad shrugged and then sighed. "She'll be a lot better when I can tell her you're alive."

Dan stared at Chad. "She thinks I'm dead?" Dan grabbed the bars in front of him. "They told her that didn't they? Those sons a bitches."

Chad nodded, he didn't need Dan to tell him who he meant when he said they. Chad stepped closer. "Is there anyone else down here?"

Hank stepped forward and answered. "Just Dan and I. A guard comes down four times a day. He brings our meals and then checks us once at night."

Chad looked behind him and then back at Hank and Dan. "When?"

Dan shrugged. "I don't know, he'll probably be around in an hour or so."

Chad nodded. "I need to get out of here before he gets back."

Hank stared at him. "What, you're leaving?"

His voice sounded way too loud for Chad. "Settle down, I wasn't even prepared to actually find you here, let alone try and make some kind of rescue. I haven't got any kind of plan to get you out."

Chad pointed at the old lock on the prison door. "Like that for starters." Chad shook his head. "Plus I need to come after the guard does his last check. I can't possibly try anything tonight. I need to go back to town and get some equipment." Chad was silent a moment and then he sighed. "A gun might not hurt either."

Dan turned to Hank. "He's right Hank. We've been here this long; I think we can give him a day."
Dan looked at Chad. "You better get out of here."

Chad nodded. "I'll be back tomorrow night, probably about midnight." Chad looked at the two men. "I'll get you out."

Dan nodded. "Would you tell Mary I love her?"

Chad smiled. "It would be my pleasure."

Chad turned and left the two men who stared at the spot he had been standing in, like he had been a ghost who had just made a brief visitation and then eerily disappeared.

Chad walked back down the hallway and into the first room. His head whipping around, watching for the guards, as he retraced his steps. Chad walked over to the spool. He hoped the guards wouldn't notice that it had been pushed under the window. Chad shrugged; he couldn't do anything about that, except hope for the best. Chad climbed on to the spool and climbed back out the window, carefully pulling it shut behind him. He hurried back to where he had left his car and then drove home and went inside.

Chad didn't relax until he was sitting on his couch. When he finally did he could feel the pent up adrenal of the last few hours finally seeping from his body. Chad drew a shaky breath and wiped a hand across his face. He had a hell of a lot of planning to do.

Chapter 10

Chad woke early the next morning and immediately his mind started ticking off all the things he needed to do today. First things first though, he went to make coffee. Chad knew without it he wouldn't even make it to step one.

With two cups finished, Chad pulled out his cell phone and called J.C.'s private number.

J.C. answered before the first ring had finished. "What's going on Chad? Are you okay?"

Chad took a breath, he was sure J.C. wasn't going to buy his story. "J.C., I'm sorry but I'm not coming in today. I don't know if it's the flu or maybe I ate some bad food, but you know it's one of those things that make you feel like you better stay home and close to the bathroom. I really hate to miss a day you know. I really am sorry."

In his office, J.C. was shaking his head. "I'm okay with you missing a day Chad, but I don't believe for a minute that you're sick. Do you want to tell me what's really going on?"

Chad wished this conversation would have went in a different direction, but he had also known all along J.C. wouldn't believe his being sick story. "Really J.C., I'm just sick. Everything else is okay, but listen if I need you I have my phone. I won't hesitate to call."

J.C. frowned. "If that's the way you want it, just remember I'll help if I can."

Chad nodded. "I know that and believe me I'm grateful. I'll try to be in tomorrow morning."

Chad hung up the phone wishing he could have told J.C. what he was up to. He shook his head and finished his coffee. It was better this way. J.C. had too much to lose.

Chad used his phone to look up a number and dialed it, glad he had a platinum visa with a hefty available balance. When the call was done, Chad left the house and headed to the hardware store. The only thing he could think of for the lock that had Dan and Hank imprisoned was a set of bolt cutters. He supposed a lock picking kit would have been nice, but wouldn't do him much good. He had no idea how to use one and didn't have time now to learn. Chad found and bought the bolt cutters and carried them to his car. He threw them in his trunk. The cutters were fairly heavy and just over three feet long, but he didn't see a problem carrying them the half mile he'd be walking tonight.

Chad was a little nervous about the next stop he planned. He used the map on his phone to find the house he wanted. He pulled up in front and studied the house. It was definitely empty, he hadn't expected any different. The lawn was overgrown, but since they hadn't quite reached summer yet, the grass was still fairly green.

The blue house Chad stared at had an abandoned feeling to it, which made sense, but it still made Chad feel empty and more than a little mad. Sitting in his car, Chad surveyed the neighborhood. Thankfully he didn't see anyone outside. He figured this was probably more of a working class area and everyone was away at their jobs. That was at least one plus on his side. Stepping out of the car Chad hoped his luck held out.

Chad made his way to the back of the house looking around as he went. If anyone did see him he knew they'd figure he was up to something just by his paranoid behavior. As he reached the back side of the house, Chad went up to the door. He reached above the door frame and then under the mat that said welcome, hoping to find a spare key. On a window sill to one side he saw two clay pots. Chad breathed a sigh of relief when he found a door key under one.

Chad opened the door and quickly stepped inside. He walked through the quiet house room by room until he found the one he needed.

Chad hated the eerie silence that permeated the house. He kept waiting for the slamming open of the back door and someone screaming about a thief in the house. Chad shook off the thought and went to a closet and found two suitcases. He loaded them up with what he needed and carried them to the back door. Chad stuck his head out and looked around. When he determined that the coast was clear he stepped out with the cases. He re locked the door and put the key back in its hiding place. Chad made a mad dash to the car and threw the suitcases in the trunk with the bolt cutters. Chad sat for a minute in his car letting his nerves settle, and then he drove home. Going inside, the first thing Chad did was warm a cup of coffee in his microwave. Then he went over the rest of his plans in his head. He just had one more important stop to make and by then it would be time to go see Mary.

Chad cooked himself a frozen meal in the microwave and ate it, feeling this might be his last chance for food today. Before leaving the house, Chad went to the safe in his bedroom and grabbed his pistol and some extra shells. He hardly ever used the thirty eight. He'd bought it when he'd been receiving some threatening phone calls a few years back. He knew how to shoot it, but thankfully had never had to use it. Chad put the pistol in his backpack with his recorder and notebooks and then reached in to the safe and grabbed a large envelope. He added that to his pack. He carried it all out to the car and locked it up with the rest of his collections. Chad drove from his house to a small shop just off the town's main street and went inside. The chimes above the door rang out when he entered. He heard a high, feminine voice holler from somewhere toward the back of the store. "I'll be right with you."

A few seconds later a short woman with tons of blonde hair came walking out. Chad broke in to a huge smile. Looking at Yvonne was like looking at a Dolly Parton look alike. She had the curvy figure to go along with the blonde hair and even a slight country twang to her voice. "Chad, for goodness sales, I haven't seen you in ages. C'mon over here and give me some sugar."

Chad stepped over and pulled the woman in to a big hug.
"Yvonne, it's good to see you, how's business?"

Yvonne stepped back and smiled. "Couldn't be better.
The prom is coming up and the girls just love my
dresses."

Chad nodded, Yvonne ran a dress and costume shop.
She had just about everything under the sun in her store.
She was the first one Chad thought of when he had
gotten his idea. He looked at Yvonne. "I need a big
favor."

Yvonne smiled. "Chad you know all you need to do is
ask. I never could resist that handsome face of yours."

Chad laughed at that, but walked out an hour later with
everything he needed. He drove to the outskirts of town
and parked. He wanted to rearrange things in his trunk,
but also wanted to avoid prying eyes. When he was sure
the coast was clear, Chad got out of his car and went
around to the trunk. He opened the lid and got busy.
First he put his gun under the cover that hid his spare
tire. Chad pushed the pistol down in where it wouldn't
show. Then he removed his recorder and notebooks
from his pack along with the envelope he had gotten
from his safe.

He laid them to one side on the floor of the trunk. Chad
retrieved the things he had gotten from Yvonne and
placed them in the now empty backpack. He picked up
the pack and examined it to make sure it didn't look too

bulky. When he was satisfied, Chad closed the trunk and carried his backpack to the front of his car and laid it on the passenger seat.

Chad got in the car and looked at his watch, five thirty. Chad nodded; by the time he got to Rose Hills it would be about time to visit Mary. Chad smiled, so far, so good. Although he knew the hardest parts of his plan were still ahead of him. The least of which was explaining to Mary about Dan.

Chad pulled around to the back side of the hospital and parked. Then he walked around to the front of the building and entered. A few minutes later he was standing in Mary's doorway. She'd been sitting at her table trying not to stare at the door in anticipation of Chad's visit. Mary still couldn't shake the feeling that one day Chad just wouldn't show up and she'd be alone again. She wasn't sure if she could take that. She smiled at him standing there. Chad returned the smile and walked over to the table.

"It looks like you may have had yourself a better day."

Mary nodded. "No problems other than being in here of course."

Chad nodded and sat sown, sliding his backpack under the table as he did. He looked at Mary. "I have something I want to talk to you about. It's a little hard to explain and probably a lot harder to believe."

Mary frowned, wondering what was going on. Then she gave Chad a small smile. "After all I've been through I'll believe a lot of things I wouldn't have before."

Chad laughed. "I guess that's true."
Chad took a deep breath before continuing.

"Okay, first off I should tell you I have been doing some digging in to Senator Buchanan. I should say someone else has been doing the digging for me. He's a guy who's been known to find out things that no one else can. I have to admit he doesn't always use what could be called the legal route. Sometimes it's not even the moral route. I have to give him credit though, he gets the job done. Besides what I learned from him, I also received another one of those strange phone calls like before."

Mary's forehead was furrowed as she tried to figure out what Chad's talk was leading to. He had to smile at the look on her face. "I'm sorry Mary, I know I sound a little crazy here, but there is a method to my madness."

Mary nodded slowly. "I'm sure there is, maybe you should just say whatever it is you're trying to get at straight out."

Taking Mary's hand, Chad nodded. "You're right; I just need to say it." Chad stared in to Mary's eyes. "Dan's alive Mary."

Mary stared at Chad for a couple of seconds, then she pulled her hand away. "What's wrong with you? What kind of sick joke is this? I thought you were my friend."

Mary couldn't say anymore, instead she just glared at Chad. He could see not only anger, but hurt in those baby blue eyes and hated that he had been the one to put it there.

"I am your friend. Do you really think I'd joke about something like this? Damn it Mary, I've seen him, I talked to him. Before I left the last words Dan said to me were tell Mary I love her."

Mary was taking deep breaths, trying to calm herself. "Why isn't he here then? If you talked to him, why didn't he come with you?"

Chad looked down at his lap. "I had to leave him." He looked back up at Mary. "They've got him and another guy locked up in this make shift jail cell out at the old mines. I found out Senator Buchanan is working with the Chinese to open the mines. Then I got the second phone call and the guy told me to check the mines. I went out there last night, I didn't really believe any of it, but I had to see for myself. Believe me; I was as shocked as you."

Mary could see Chad was telling the truth. She should have known better than to think he'd lie about something like this. After all she had been through it was hard to trust anyone any more.

Mary shook her head. "I'm sorry Chad, sorry I doubted you."

Chad smiled. "It is a little hard to wrap your head around."

Mary leaned toward Chad. "What are you gonna do now?"

Chad looked toward the door and then lowered his voice as he turned back to Mary. "I have a plan." Chad reached under the table and pulled his backpack out. He sat it on the table. "This may sound a little crazy." He laughed. "More than a little I guess, but I think we can pull it off."
Chad opened the pack. "I have a friend that owns a costume shop. I told her you were going to a masquerade party and needed something to change your looks so no one would recognize you. Something that didn't include a mask."

Mary was frowning. "I don't think I'm following you Chad. Why do I need a costume?"

Chad smiled. "Because you and I are both going to be walking out of here tonight."

Mary stared at Chad like he was out of his mind, which she was pretty sure was the case. Then she shook her head. "No one is going to let me walk out of here."

Chad laughed. "You won't be you." He held out the backpack to her. "You go in the bathroom and just try this on. I promise, you won't even recognize yourself."

Mary just stared at the bag. Chad pushed it toward her. "Go ahead, take it. Try on the costume and see what you think at least."

Mary nodded. "Okay Chad, but I'm telling you it won't work."

Chad sat back in his chair and folded his arms. He had a big grin on his face. "We'll see."

Shaking her head, Mary took the backpack and headed for the bathroom. At the doorway she turned back to look at Chad. She couldn't help but smile at the look of confidence on his face. Mary walked into the bathroom, closing the door behind her. The room she stepped into was fairly barren. Mary had never worried about bringing any personal belongings with her, not that she'd been given much choice in the matter. The bathroom consisted of a tub with a shower, a toilet and a small vanity sink with a mirror over it. Mary set the backpack down on the vanity top and looked inside. On the top was a blonde, curly wig. She pulled it out smiling. It was definitely different than her own straight brown, shoulder length hair. Below the wig Mary reached in and found what she thought was a dark blue dress. As she pulled it out she realized how wrong she was. The whole dress was padded. At least two inches of thick padding lined the inside of the dress. As Mary looked more closely she noticed the padding in the chest area was at least double the padding elsewhere. She figured with the dress on she would look about forty pounds heavier. Mary removed her outer clothing and slipped in to the dress, surprised at how well it fit. She looked in the mirror and drew in a surprised breath. Forty pounds may have been an under estimate. Mary picked up the wig and shook it out. Placing it on her head she tucked her own dark hair up inside. Mary looked in the mirror again, her blue eyes opening wide in shock at the transformation. She shook her head and smiled. Mary looked back in the bag and found a pair of glasses and slipped them on. They weren't sunglasses, but were tinted enough to hide the blue of her eyes. Mary stared at herself for a few minutes, surprised to find herself thinking that Chad might just be right; maybe they could pull this thing off.

Mary opened the bathroom door and stepped out. Chad looked at her, his smile growing larger. He stood and walked over to her. "It's amazing Mary, you look totally different." Chad turned to look back towards the door to her room. "I think you better stay in the bathroom until I check and make sure the coast is clear. Why don't you put your other clothes in the backpack? If there is anything else you want you can throw it in there too."

Mary nodded, although she didn't have much here that she cared to keep, including a lot of bad memories. "By the way, how did you know my size? This dress fits almost perfectly."

Chad laughed. "Give Yvonne credit for that. I just gave her a description of you, she performed the magic."

Mary smiled. "Someday I'll have to meet her and thank her."

Mary stepped back in to the bathroom and closed the door. Chad walked across the small room and looked out in to the hallway. It looked empty now, but he figured they probably did some kind of rounds before the patients went to bed. He wanted to be out of here and long gone with Mary before that.

He went back in to the room and went over and knocked on the bathroom door. Mary opened it. "How does it look out there?"

Chad smiled. "I didn't see anyone. Do they usually do some kind of bed check or something?"

Mary nodded. "They come around at nine o'clock every night."

Chad looked at his watch. "That gives us a couple of hours to work with. I think we had better get out of here while we can. Hand me that backpack."

Mary grabbed the bag and handed it to Chad. He took it and looked at her. "I think it will be best for you to walk out by yourself. They have seen me around. I'll follow a minute or two after you leave. Just walk out of here like you are coming from a visit. Stand up straight and walk with confidence, like you have every right to be here. No way in hell will they recognize you in that get up. Just walk out the front doors and then head around to the back parking lot. Wait for me there. Can you do that?"

Mary shrugged and then smiled. "I think so, just don't wait too long before you follow."

The two walked to the door and Chad looked out again. A few seconds later he took a hold of Mary's padded shoulder. "Okay, just go and keep walking. I'll meet you out back." Chad smiled. "And don't worry, we can do this."

Mary nodded and started walking. She made it all the way to the front doors when she heard a voice yell from behind her.

"Have a good night."

Mary's heart skipped a beat. She turned her head slightly and raised her voice to a higher pitch than normal and yelled back. "Thank you and the same to you."

Mary opened the door and stepped outside. She took a couple of deep breaths and then straightened her shoulders and forced herself to walk slowly and calmly around the building to the back parking lot. Mary frowned at the five cars she could see parked there. She wasn't sure which car was Chad's, so she walked to the far end of the lot. Mary turned so she faced the building and then waited and watched.
What seemed like an eternity passed before Mary finally sighed at the welcome sight of Chad stepping around the corner of the building.

He smiled at Mary and then pointed at his car. "Let's get out of here."

Mary just nodded. She was too wound up for speech.

Fifteen minutes later Chad pulled in to a gas station. He handed Mary the backpack. "Why don't you go in and get changed. I need to fill up the car and then we can get out of town."

With blue eyes sparkling, Mary smiled. "That's something I'll be more than happy to do. This costume itches."

Mary walked inside avoiding looking at anyone and headed straight for the gas station's bathroom. Thankfully although the station was fairly busy, the bathroom was empty. Mary changed quickly and came back out carrying the costume stuffed in Chad's backpack.

Chad was standing in line at the counter. He smiled at her as she walked toward him. "Would you like to get anything before we go?"

Mary thought for a minute. "You know I would love a mocha cappuccino, we never got anything like that at Rose Hill and we could probably use a drink for the drive."

The two went over to the coffee machines and grabbed their drinks. Chad was laughing.

"Maybe caffeine isn't such a good idea. To tell you the truth I'm really keyed up right now, but later I'll probably be grateful for the energy."

Chad paid for their drinks and the two walked out to the car. Suddenly Mary started laughing. Chad looked at her. "What's so funny?"

Mary shook her head. "I was just thinking, I don't think anyone noticed me when I walked in, but if they did I can't imagine what they thought. I walked in a heavy set blonde and then in a matter of minutes I walked out with not only a new hair color but a totally different body size as well."

Chad joined in the laughter, it felt good after the nerve racking ordeal of getting Mary safely out of Rose Hill. "If you could patent your technique for that you would be a millionaire. Maybe not the hair, but that is some fast weight loss."

The two got in Chad's car and he headed south of town. When Mary finally sobered up she looked over at Chad. "Do you have a plan for us to get Dan out of that place?"

Chad glanced at Mary, frowning, before focusing on the road again. "I have a plan for me to get Dan out; you are staying in the car."

Mary frowned now. "I don't want to stay in the car; I want to help get Dan out."

Shaking his head, Chad answered back, his tone firm. "You're staying in the car Mary. It's going to be hard enough for me to get in there and then try to get back out with not only Dan, but that Hank guy too. Besides if something does happen, I want you here, in the car with my cell phone and my boss, J.C.'s number on speed dial."

Mary folded her arms across her chest and looked out the window. She knew Chad was right, but she wanted to see Dan. For her, Dan had been dead for over six months and now it was like he was being resurrected and miraculously brought back to her. Until she saw him though, it was still like she was dreaming.

Chad turned off the main road. "You better hang on; this road is a little rough."

Mary pulled herself out of her reverie to look ahead of them. It was starting to get dark, but there was enough light that you could still see well enough to make out several big ruts in the road ahead. Chad looked at Mary quickly and then back at the maze he was trying to drive through. "I know it's kind of hard to see, but I don't want to turn on my lights. I'm not sure who might be around."

Just then the car hit a large pothole and Mary bounced in the seat, her head less than an inch from hitting the ceiling. She was laughing. "Hey, I'd rather be bounced around a little than be caught and sent back to that hospital."

Smiling, Chad nodded. "I'll try to keep the bumps and bruises to a minimum anyway."

Mary smiled back. "You're doing fine; in fact I haven't even thanked you yet for getting me out of that place."

Chad pulled over to the side of the ride and stopped the car. "You don't have to thank me Mary. All I did was the right thing. God knows you never should have been put in that place or been put through the hell of thinking Dan was dead. I hope someday to write your story and make Senator Buchanan and whoever else was behind the suppression of Dan's stack and the screwing up of your life pay for what they did."

Mary frowned. "What do you mean some day? I thought you were going to be publishing the story right away?"

Chad shook his head. "I didn't want to tell you or have the people at Rose Hill find out, but J.C. ordered me not to do the story."

Mary's face scrunched up in anger. "Why the hell would he do that?"

Chad held up his hands. "Wait a minute, don't be mad at J.C., someone threatened him if that story ever comes out. I'm sure he would have lost the paper and that's just for starters. Believe me, that paper is his life. I could have never asked him to even take the chance of losing that."

Chad looked at Mary's face. "Don't worry, one way or another I'll write the story. Maybe not for J.C.'s paper, but your story needs to be told. As long as I have your

155

permission and Dan's too, when I get him out, then the story will come out."

Mary smiled. "You have my permission for sure and I know Dan will gladly agree."

Mary was silent a moment and looked upset. Chad frowned. "What's the matter?'

Mary shrugged. "I'm just worried after you get Dan out, what do we do then? The same people that did all of this to us aren't just going to stand back and let us walk away. They will be after both of us and probably you too and who knows what they'll do?"

Chad grinned. "Don't worry, I have a plan."

Mary laughed and her voice sounded like music in the otherwise quiet car. "If you say so then I believe you. After all you've done already; I'm beginning to think you're a magic man."

Chad laughter filled the car. "Hardly and I don't think we should push our luck. Let's not make me into a magic man until I have Dan safe and sound and sitting in the car."

Mary nodded, but she already had Dan free and with her. Mary's mind wouldn't let her think any other way.

Looking at his watch, Chad shrugged. "It's just a little after ten. I'm going to wait until eleven thirty and then start walking." Chad pointed out through the front window of the car.

"About a half a mile up that way is the back of the main building. That's where I'll be headed."

He turned to Mary. "You are going to sit here. I want you to give me two hours before you even start worrying. If I'm not back by then I want you to call J.C. I really hate to involve him, but if for some reason I don't make it back here, you make the call. He'll come running and I'm pretty sure he'll be bringing some kind of back up. No matter what happens though I don't want you leaving this car." Chad stared at Mary. "I mean that and I want you to promise me you won't."

Mary just stared at Chad, but she knew by the look on his face she had better not argue. "Okay Chad, I promise I won't leave the car. If something happens, if for some reason you don't come back within two hours I'll call your boss."

Chad nodded. "And you'll stay in the car then too, J.C. will handle it."

Shaking her head wearily, Mary finally nodded. "I promise Chad, I'll stay put."

Smiling at her, Chad nodded. "Good, now we have over an hour to kill, what do you want to talk about?"

Smiling, Mary laughed. "Well, you know all about me. How about we talk about you for a while."

Chad laughed. "What do you want to know?"

Mary tipped her head thinking. "How about why you became a reporter to start with."

Chad shrugged. "Ever since I was a little boy I loved telling stories. I remember in fourth grade we got to write stories for a make believe newspaper. All of us kids wrote the stories and our teacher typed them up and then printed them out so they looked like a real paper. My story got the most votes as being the best liked. I think that way back then is when the bug first bit me."

Mary smiled trying to imagine what a nine or ten year old Chad must have been like. She listened as Chad talked, mesmerized by his stories. She was surprised when Chad looked at his watch and shook his head. "I guess it's about time."

Mary's eyes widened. "Really, it feels like we've only been talking a few minutes." Then she frowned. "You be careful Chad. I want to see Dan, but I also want to see you safe and sound just as much."

Chad laughed. "That goes for both of us."

He beat his fingers on the steering wheel a couple of times, thinking. Then he handed Mary his cell phone. "All you have to do is hit the number two and then send. That's J.C.'s number."

Mary smiled. "Who's number one on there?"

Looking at Mary, Chad smiled. "Actually, that's my Dad. We're pretty close. My mom died when I was a kid and my Dad raised me, so he's number one on my phone and in my life."

Mary shook her head. "I'm so sorry Chad."

Chad shrugged. "It was a long time ago and personally I think my Dad did a terrific job."

Mary smiled. "I think so too."

Chad took a breath. "I'm going to get a couple things from the trunk and then I better be heading out."

Mary nodded and Chad got out of the car and went around to the back. He opened the trunk and dug down in the spare tire compartment and retrieved his gun. He tucked it in at the back of his waist band and then untucked his shirt and pulled it over the top of the gun to hide it. He didn't want Mary to see it and worry more than she already was. Chad then grabbed out the bolt cutters. He shut the trunk quietly and went back around and opened his car door. He looked at Mary and then pointed at the glove box. "Could you grab me the flashlight out of there?"

Mary opened the compartment and handed Chad the flashlight. She held up Chad's phone. "It's eleven thirty seven; you have two hours from now."

Chad nodded. "I plan on being back with Dan and Hank and driving out of here before then."

Mary nodded and tried to smile. "I'll be waiting."
Mary watched Chad walk away and tried to still her
heart that felt like it was going to beat right out of her
chest.

As Chad walked away, he had to force himself not to
look back. He knew if he did it would be too hard to
keep walking. He focused on reaching the mine building
instead. Twenty minutes of rugged walking and tripping
over what he figured must have been every stone in the
county; Chad stepped up behind the building. It looked
almost the same as it had the night before. Although
Chad did notice that less light was coming through the
few boarded up windows and the cracks in the old wood.
Chad hoped that meant the guards or whatever they were
had gone to sleep.

Chad hurried around to the basement window he had
used the night before. He held his breath as he pushed
on the window. He was sure that he'd been found out
and the window would be locked when he got here.
Chad let out a sigh of relief when he was able to easily
push in the window.

He leaned in through the window and used the handle on
the bolt cutters to hang them on the hook that was just
inside the window on the wall. Then he swung himself
in and dropped quietly to the floor. He took down the
bolt cutters and then put his hand on the small of his
back to reassure himself that his pistol was still there.
Chad stood in the room for a minute and listened. He
didn't hear anything and hope that was a good sign. The
idea flashed through his mind that he would step in to

the other room only to find that the cell in there was empty. Chad shook the thought from his head and walked across the room and in to the hallway.

He walked down the hallway almost sideways, trying to watch in front of him and behind him at the same time.

Chad stepped in to the room at the end of the hall and sighed with relief when he saw Dan and Hank standing in the cage. Both men stared at Chad. Chad glanced behind him one more time and then hurried over to the jail cell. He looked from Dan to Hank. "Man, am I glad you're both still here."

Hank grunted. "That's a hell of a greeting."

Chad smiled. "Sorry, I just kept thinking I'd get here and find they had moved you."

Dan smiled. "Since we are still here, maybe you could use those things you're holding and get us out."

Holding up the bolt cutters, Chad nodded. "Yeah, right, just give me a minute."

Chad put the bolt cutters on the lock's hasp and squeezed. He felt like his arms would break before the damn lock did. Finally when he felt like nothing was going to happen, the cutters split through the metal and the lock broke, clanging loudly as it landed on the cement floor. Chad looked quickly behind him, and then he turned back and pulled open the door. "Jesus, that was loud, we better get moving before someone comes to investigate."

Chad didn't have to say that twice. Both Dan and Hank stepped out. Dan smiled. "Thanks Chad, now which way do we go?"

Chad pointed to the hallway. "Follow me, there's another room just through here."

Chad headed down the hall with Dan and Hank close on his heels. On the way out Hank grabbed the bolt cutters. "If we get caught, I'm not going down without a fight. I think these things could do some damage."

Chad thought about the pistol he had concealed, but decided not to mention it. The three men walked through the hall and in to the other room. Chad pointed over to where the large spool sat beneath the window. "We can climb out over there."

The trio hurried over. Dan pointed at Hank. "You go up first, I'll hold those cutters until you get out."

Hank was going to argue about who should go first, but didn't want to waste any time. He nodded instead, handed Dan the bolt cutters and then jumped up on to the spool. Hank quickly made it out the window and then reached back in and retrieved the bolt cutters from

Dan.

As soon as he handed them off, Dan turned to Chad. "Now you, get out the window."

Chad shook his head. "You first, I made a promise to Mary to get you out of here and I plan on keeping it."

Smiling at the thought of his wife, Dan nodded. "We wouldn't want to break that promise, but I want you close enough behind me that I can hear you breathe."

Chad shook his head. "Hey, no problem there."

Dan climbed on to the stool and out the window with Chad right behind him.

When the three had gotten out, Chad pointed. "We have to go that way; my car is about a half mile away."

Chad started walking as fast as he could without making noise. Dan and Hank stayed right behind him. The three walked around to the back of the building and started heading for the trees when they heard the door at the back of the mine building slam open. Someone started shouting in what Chad was certain had to be Chinese. He looked at Dan and Hank. "We're going to have to make a run for it, follow me." The two nodded and all three men started running for the trees.

* * *

In the car Mary looked at the digital clock on Chad's cell phone. It said twelve nineteen. Chad had been gone forty two minutes. He should be there and inside by now. Mary closed her eyes and took a breath. Please, be okay, please, be okay. Mary kept repeating the words in her mind.

Her closed eyes flew open when she heard the first shot.

* * *

As the three men heard the shots, Chad was trying to push Dan and Hank ahead of him.
"Keep running, it's straight ahead now."

Chad reached behind him to the back of his waist band and pulled out his thirty eight.
"I'll try to hold them off, you two get to the car."

Dan shook his head. "We're not leaving you, just try firing a few shots back at them, maybe you can buy us all enough time to get to the car.'

Chad pointed the gun back the way they had come and fired two shots. He fired another into the trees back the way the fired shots had come from.
The three ducked as an answered shot whizzed by just above their heads.

The three took off running again, hunched over as they went.

As the three started off they saw a car headed toward them, the lights off. Chad shouted.
"That's my car."

Dan looked at him. "How can it be your car? Who the hell is driving it?"

Chad kept running, but he was smiling.
"Mary's driving, God bless her."

Mary stomped on the brakes and turned the wheel sharply, making the car spin sideways in front of the men.

Mary threw the car in park and jumped out, leaving the engine running. She ran straight to Dan and was wrapped in his open arms.

Hank stepped past them. "Time for that later, get in the damn car."

Chad nodded. "He's right; Dan and Mary get in the back. Hank, you get in the passenger seat."

Chad ran over and jumped in the driver's seat. He handed his gun to Hank. "If they get close, shoot the bastards." Hank nodded as he took the gun.

Chad stepped on the gas, spun the wheel and headed out the way he and Mary had come in. He didn't bother to slow down and the car bounced and slid its way up the rutted road.

When they got to the main road, Chad slowed to ten miles above the speed limit.

He turned to the others. "I think we're okay now. I don't think they'll try to follow us in to town, at least not yet." Chad turned to Hank.
"I'm not sure what to do with you. There's no way you're gonna be able to stay around here. They'll get regrouped and then come looking for you."

Hank just laughed. "Don't worry about me; I can handle whatever they send, now that I know what they are doing. My brother has a place west of town. If you can get me there then you can wash your hands of me."

Chad looked over at Hank shaking his head.
"I don't want to wash my hands of you. Can your brother keep you safe?"

The sly smile Hank gave him spoke volumes as he nodded. "We've been through shit like this before, maybe not as intense, but we can handle it."

Chad nodded relieved. "Okay Hank, just tell me the way."

Twenty minutes later Chad pulled up a dirt lane that was well off the beaten path. A quarter mile up the road in a grove of trees Chad pulled up to a house that stood almost hidden by the trees. Hank pointed. "This would be the place." He reached over and shook Chad's hand. "You saved my life, if you ever need anything, all you have to do is ask."

Chad smiled. "I'll remember that, the way things are going I may just take you up on that offer."

Hank turned to the couple in the back seat. "You two take care of each other."

Hank reached over the seat and shook Dan's hand, hanging on to it tightly for a moment. "I couldn't have been locked up with a better person."

Dan smiled. "The feelings mutual, stay safe Hank."

Hank nodded. "You too, Dan." He turned to Mary. "That goes for you too Mary. Dan speaks highly of you and you know what?"

Mary shook her head and Hank smiled. "I believe every damn word."

Hank got out of the car, waved and then disappeared in to the trees. Chad turned the car around and headed back towards town, but before he got in to the city limits, he turned and took a different road.

Mary leaned forward from the back seat and grabbed Chad's shoulder. "Where are we going Chad?"

Chad just shook his head. "I have a plan remember. We'll be there in fifteen minutes. You both just sit back and relax. After all that we just went through, I bet you could use a breather."

Mary's hand tightened on Chad's shoulder. "What about you?"

Chad shook his head. "I'm okay, but when this is over I'm going to be sleeping for a week."

Dan looked out the window and smiled. He was almost certain where Chad was taking them.
Dan reached over and pulled Mary back against him. He laid his head on top of hers and breathed in the scent of his wife, the love of his life.

Mary closed her eyes as Dan's strong arms encircled her. Nothing had ever felt better.

Fifteen minutes later, Chad drove in to the airport parking lot. He pulled the car in and parked at the back end of the lot. Turning off the engine Chad turned in his seat to face Dan and Mary. He smiled and pointed at

Mary. "Do you still have my phone? It looks like I lost my watch somewhere."

Mary laughed. "It slid on to the floor up there when I skidded in to pick all of you up."

Chad laughed and bent down and retrieved his phone. He opened it looked at the clock. "Looks like we've made it with an hour to spare."

Mary frowned. "What are you talking about Chad? An hour to spare for what?"

Chad smiled at Mary, then at Dan. "I booked a flight for both of you. I heard about this little Island when I was doing a story a year ago. I think it will be the perfect place for the two of you to lay low for a while."

Mary and Dan could only stare at Chad for a moment astounded. Chad smiled at the looks.
"Oh, there is something else."
He opened the door and got of the car. Dan and Mary got out looking curiously at Chad and joined him. Chad stepped around to the back of the car and opened the trunk. "I had to go in to your house to get these. Sorry I didn't have time to get your permission, but I didn't think you'd mind."

Dan and Mary looked at their luggage sitting in the trunk. Chad reached in and grabbed the envelope he had laid in their earlier. He handed it to Dan. "There's five thousand dollars in there."

Both Dan and Mary shook their hands and started to argue, but Chad shook his head. "It's not a gift, if you'll let me, I'd like to give you that money as a down payment for the rights to your story. Personally I think it's going to be just a drop in the bucket compared to what the story will be worth. It's all I have right now though."

Mary felt tears come to her eyes. She hugged Chad. "We can never thank you enough for all you've done. You're more than a magic man, you're a miracle worker."

Chad smiled. "I think you two are the miracle workers." He turned to Dan. "So, do we have a deal?"

Dan nodded as he pulled Chad in to a hug also. "We have a deal, but only if that five thousand counts as full payment. Maybe if you tell the story I will actually be able to build the stack like I wanted."

Chad laughed. "Why don't we go in and see if we can find a cup of coffee and we can talk about it while we wait for your flight."

Chapter 11

Mary and Dan sat side by side on lounge chairs. They
were watching four small children playing in the ocean's
blue water. Both were laughing at the children's antics.
The kids would scream and run for the safety of the
sandy beach each time a small wave would crash toward
the beach front.

Dan reached over and grabbed Mary's hand.
"Are you happy here Mary?"

Smiling, Mary nodded. "I am, mostly because I'm with
you. Chad gave me the best gift in the world. From the
day he walked in to my room at Rose Hill Psychiatric
Hospital Chad gave me hope. Even before we knew that
you were alive. He believed in me and no one had done
that for a long time. I have to say I miss him a lot.
There's not much else I miss though. Both are parents
are gone."

Dan nodded. "He'll come visit when he can."
Dan looked in to Mary's eyes. "So if I asked you to stay
here and never go back you wouldn't be upset?"

Mary laughed. "Are you kidding, these last three months
have been like a dream come true."

Mary smiled at Dan. "I can't tell you how bad I missed
you. I thought I'd never see you…"
Mary couldn't finish the words as she began to cry.

Dan sat up in his chair and pulled her in to a hug. "It's okay Mary, the nightmares are over now."

Mary sat back. "What about your invention?"

Dan shrugged. "Someday the time will be right. I can work on that anywhere though, might as well be in paradise."

Mary smiled and her face lit up. "I guess we're going to be Islanders then. I hope Chad visits soon so he can see what wonderful things he's done for us."

Dan nodded and the two went back to watching the children.

Chapter 12

Chad sat as his typewriter. Things had finally settled down since Mary and Dan's 'disappearance' and it was time.

The story was really already written; he just had to put the words on paper or rather on the computer. Chad thought about Mary and Dan. He barely knew them and yet he felt he really knew them as well as he knew himself.

Chad shook his head and sighed. It was time, time to tell the world their story. He began to type.

Mary was at the stove cooking dinner when she heard the back door open. She looked up at the clock and smiled. "I'm in the kitchen Dan.

This book is dedicated to the many inventors who have inventions and ideas suppressed by people with too much money and too much power.

It is also dedicated to the real "Dan" who had the invention in this novel patented years ago and despite his best efforts was blocked and suppressed with each step he took. The patent still sits in the patent office waiting to come to light to help save this earth we all share.

Hopefully. Some day people will see what is going on and why the government doesn't want free energy or a car that gets over forty miles per gallon. The people will eventually be heard and then we can have cleaner air, cheaper energy and gas.

To all those garage inventors, I want to say thank you and always believe!

A special shout out to the readers who have made this and the other P.S. Winn books possible.

It is you and your imaginations that finish the stories and I am grateful.

As always, thanks to my family and friends for your patience and support, especially my husband who knows what this story is truly about.

P.S. Winn

Author bio and books available on Amazon and on Barnes and Noble.
Just look up 'P.S. Winn' for books for all ages and genres. Thanks all for the great support in the crazy journey I call being an author.